Principles of
Functional Verification

Principles of Functional Verification

by Andreas Meyer

ELSEVIER

AMSTERDAM • BOSTON • HEIDELBERG • LONDON
NEW YORK • OXFORD • PARIS • SAN DIEGO
SAN FRANCISCO • SINGAPORE • SYDNEY • TOKYO

Newnes

Library of Congress Cataloging-in-Publication Data

Meyer, Andreas.
    Principles of functional verification / Andreas Meyer.
        p. cm.
    ISBN 0-7506-7617-5
    1. Integrated circuits–Verification. I. Title.

TK7885.7.M49 2003
621.3815′48–dc21

                                                      2003056421

British Library Cataloguing-in-Publication Data
A catalogue record for this book is available from the British Library.

The publisher offers special discounts on bulk orders of this book.
For information, please contact:

Manager of Special Sales
Elsevier Science
200 Wheeler Road
Burlington, MA 01803
Tel: 781-313-4700
Fax: 781-313-4880

For information on all Newnes publications available, contact our World Wide Web home page at: http://www.newnespress.com

10 9 8 7 6 5 4 3 2 1

Printed in the United States of America

Contents

Preface

Functional verification has become a major component of digital design projects. Various industry surveys show that verification is the single largest component of the project, taking up more than half of the total project's staffing, schedule, and cost. It is often the limiting factor to project completion, and is becoming the single largest bottleneck in the industry.

Functional verification is complex, time-consuming, and sometimes poorly understood. As a result, the verification effort represents one of the bigger risks to the successful completion of a project.

By providing guidelines and insight into approaches that have been successful in real-world projects, this book details sets of methods, tools, and disciplines to smoothly and effectively verify a project.

The examples used in the book are based on verification techniques that have been used in actual, successful projects. Because these examples are fairly complex, they are treated in an abstract fashion to show the intent. There is no one way or one language that works best for functional verification. The book is language neutral, and shows the strengths and weaknesses of various approaches so that a verification plan can be created based on a broad range of techniques and the requirements of the specific project.

For someone with no previous experience in functional verification, the book introduces the concepts and issues in the field, assuming a basic understanding of hardware design and simulation.

For a project manager, the book will also provide an understanding of the time and risk tradeoffs, as well as how to plan, integrate, and manage the verification portion of the project.

For an engineer who is already involved in functional verification this book may provide new insight and methods to improve the verification process.

Functional verification will continue to be the largest piece of a project. This book is intended to help engineers manage and understand the verification process so they can reach the goals of predictable flow, efficient use of resources, and a successful project outcome.

Acknowledgments

I would like to thank Robert Fredieu, and my wife, Renee Le Verrier, for their detailed reviews, reorganizations, copy edits, and support.

Introduction

Verification of electronic designs has become common as the number and complexity of designs has increased. Common industry estimates are that functional verification takes approximately 70% of the total effort on a project, and that the verification process is the single largest issue in completing a new project.

There are many different types of verification that are used when designing or building a new system. These may include manufacturing verification, functional verification, and timing verification, among others. Each has specific goals, and is used for very different tasks.

Perhaps the oldest verification is called manufacturing verification, or verification test. This is where a manufactured component, perhaps an application-specific integrated circuit (ASIC) or a board is verified after it has been manufactured. The purpose of this type of verification is to ensure that a single specific part has been manufactured just like all other parts. It meets the specifications of the design, and it operates exactly like the first or any subsequent part. It can be viewed as an equivalency test that determines if any one particular part is the same as the first one that was built. Manufacturing tests are usually run on every single device that is built, and look for identical behavior between devices.

Unlike a manufacturing test that is run on every part that is built, functional verification focuses on the design before a single part is built. Functional verification attempts to determine if the design will operate as specified. This requires a specification that indicates what constitutes correct operation, or how the device is intended to function.

This area of intent—looking at the design and implementation of the system and components before they have been built, and perhaps the first few after they have been built—is the focus of functional verification.

Why Functional Verification Is Needed

Historically, functional verification was done once a prototype component was built. The prototype was tested to see if the design implemented the intent of the architecture by plugging it in and trying it out. If problems were found, then the prototype was modified as necessary to fix the design.

For this method to work, a few conditions had to hold true. First, it had to be possible to observe the internal operation of the prototype to determine where a problem occurred. Second, it had to be possible to modify the prototype to correct the problem. Third, one had to expect that the total number of problems would be small enough that they could all be reasonably corrected in the prototype.

It is rare for systems designed today to meet all of these conditions. It is more likely that a project will meet none of them. Few designs are built of discrete components anymore. That can make it difficult to modify points in a design that are buried inside a large device. Modifications between components are also more difficult. In addition to the complexity of modifying some larger components, the boards that hold the components now rarely have only two or four layers of wires, and they tend to be tuned to pass high-speed signals. Cutting or replacing wires on a board has become difficult, and would result in impedance mismatches and crosstalk that would result in new failures.

In addition to the problems with modifying boards, as ASICs and systems on a chip (SoCs) have become commonplace, the ability to observe the design disappears, and the cost of making a modification is measured in millions of dollars. Even field programmable gate arrays (FPGAs), which are easier to modify, will hold on the order of a million gates in a single device, preventing any reasonable measure of observability.

Finally, with design sizes measured in millions of gates, one can expect to find hundreds of errors in the initial design. This is generally too large a number to be practical to fix in a prototype.

As a result, most modern designs need a testing method that can be used before a prototype is built. This is the purpose of functional verification.

The Goal of Functional Verification

The goal of functional verification can be simply stated: to prove that a design will work as intended. There are four components to achieving this goal: determine what the intent is, determine what the design does, compare the two to ensure that they match, and estimate the level of confidence of the verification effort.

Determining Intent

Determining the intent of the system is a necessity for functional verification to succeed. The intent could be defined as what the system is supposed to do, which may be different from what it actually does. In some cases, the intent may be obvious, when it is possible to clearly specify the functions that the device must perform. However, for most reasonably complex digital systems even the intent may not always be clear. How should a device operate when an error is introduced, or when two competing actions are received at the same time, or when a resource is oversubscribed? It is often the architectural specification that describes the intent of the system and the implementation specification that documents how the design is intended to implement the intent.

Architectural specifications are usually created by examining many use-cases, which are specific scenarios that describe how the device will work, and determining the system intent for each use-case. An architecture is then defined to satisfy all of the use-cases. This is a reasonable, common approach. The problem that occurs for complex systems is that one can't consider every possible scenario of events, and as a result, the architecture will not always meet the intent of the system.

In some cases, it may be sufficient to verify that the design accurately matches the specification. However, the specification itself may have mistakes. In most complex projects, there are generally a significant number of specification errors. That is not surprising if the design specification was created manually from the architectural documents. The same may be true of the architectural specification. Not all use-cases will have been considered, or there may be discrepancies between use-cases. Part of the process of functional verification is to explore conditions that have not been considered by the architectural specification, or do not match the intent of the system. All of these should be found and corrected during the course of functional verification.

Generally there is a hierarchy of specifications that is created and used as a project is underway. While many different names are used for various specifications, Figure 1-1 shows the nomenclature that is used in this book.

There are generally one or more documents that specify the architectural requirements of a new document. From these, an

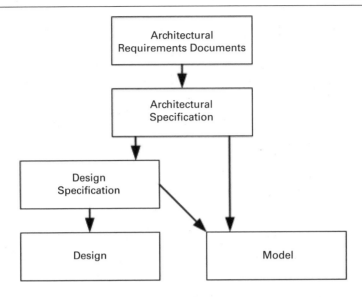

Figure 1-1: Specification Hierarchy

architectural specification is created. This document generally specifies the functions and components that will be needed to create the new design.

From an architectural specification, a design specification is created that outlines how the architectural requirements will be implemented. With the design specification complete, a design will be created.

The model of the design is project dependent. In some cases, the model may simply be the implementation of the design. In other cases, a separate model will be created. The purpose of a separate model varies based on project requirements. Some uses of models are discussed in later chapters.

Determining What a Design Does

The second part of functional verification is to determine what a design does. This is the first step in comparing the design with the intent. Since the design has not yet been implemented in a prototype, a model of the design is needed. The model is built in a software language that allows simulations to be run. The method used to test the model is quite traditional: poke it and see what happens. The poking is frequently referred to as stimulus injection, while the observation involves collecting and checking the outputs of the simulation.

Most systems are able to run many different types of operations, often in parallel. In order to fully determine what a design does, it is important to examine each operation that is supported. Parallel operations dramatically increase the complexity of this task, since ideally all possible combinations of operations must be examined to determine how the system will react. This is one of the reasons that functional verification can become quite complex.

Comparing the Intent with the Design

Ideally, one would like to take the intent and the design, poke them the same way, and check that the results match. That would provide a direct way to validate that the two are identical. The issue with this is that intent is not something that can usually be modeled. In some cases, it may not even be fully understood. Instead, something else must be used to represent the intent of the system. In some cases, people have built executable design specifications. That can be useful, but of course, the specification may also have errors. In other cases, the intent is captured as a reference model that is supposed to behave as the system is intended to work. These are some of the uses of the model block shown in Figure 1-1. Still another is to provide a series of tests and expected results to run against a design. These approaches, and most others, rely on comparing two different models, and examining the discrepancies. It is hoped that two models will not have exactly the same errors.

In all of these cases, the challenge is in representing the intent of the system. While the intent may be well understood in the minds of the architects, that information needs to be extracted into a model that can be examined in a simulation. Another concern is that the intent is rarely defined completely for reasonably complex systems. There are usually situations that were not considered. When these situations are discovered, the architecture is modified to include the new situations, and both models must be updated accordingly.

Determining Completeness

There are a number of other complexities that arise during the process of functional verification. One of the most frequently discussed is determining the completeness criteria of the verification.

Just as the system intent is often not fully defined, the design model may never be fully tested. There are several reasons for this. First, it is difficult to ensure that the complete intent of the system is known. Anything that was not included in the architecture or functional specification is unlikely to show up in the design or be tested, even though the function may be required for the system to work. Second, it is difficult to know if one has poked at the design sufficiently, and asked all the important questions. Any issue that is found provides proof that the questions were not all previously asked. However, it is rare that one can prove that no other questions need asking. This is a fundamental problem with a negative proof. One generally can't show that there are no more errors to be found. Another limit to functional verification is the impracticality of running a complete test. Any reasonably sized design is too big to examine completely. Ideally one would want to see that all possible stimuli are checked in all possible states of the design. Even a tiny design, with only a few hundred registers will have more states (2^{100}) than one can ever verify.

As a result, functional verification is rarely certain and rarely complete. Given the importance of success to many verification projects, estimation methods are used to provide an approximation of the quality level of the functional verification.

Organization of This Book

This book examines the principles of functional verification in more detail, and introduces a number of methods that are commonly used to solve these problems. Each chapter begins with a list of key objectives that are elaborated on throughout the chapter.

Chapters 2 and 3 explore some basic issues around modeling and architectures. Here, some of the standard methods of simulation and modeling are examined, and some common nomenclature is defined.

Chapter 4 looks at common methods of functional verification. While many of these methods can become quite complex, the goals are straightforward: inject stimulus into a system, determine how the system was intended to work, and compare the results with the design model. Structure, modularity, and re-use for functional verification are important methods to cope with the increasing complexity of designs. These chapters examine different verification structures and provide examples for various types of systems.

Chapter 5 introduces the goals and methods of randomization, and elaborates on how randomization is used in various verification methods. Randomization is a powerful tool that is an essential part of most verification processes. However, it must be used properly in order to be effective, and to avoid misleading results.

When hardware and software designs are related, co-simulation may be an effective tool for the integration and verification of both components. Chapter 6 explores the goals and requirements of hardware and software co-simulation.

Once tests have been run on the design, estimations of the completeness can be made to determine how well the design has been tested. Chapter 7 examines some of the methods for estimation. Since all of these methods are estimates, it is important to understand the assumptions, limits, and risks of the estimation techniques.

With the basic methods understood, the practical organization of a verification project is explored. Verification can be complex and time consuming, often representing a significant portion of an entire engineering project. A successful project requires a well-planned approach. Chapter 8 focuses on the verification planning and the final chapter discusses how the verification can fit into a complete project flow.

The result is an overview of the principles of functional verification, and how they are applied to a design project. Most verification projects today are using these methods to successfully verify projects.

Approach of This Book

This book discusses the methods and approaches that are commonly used in real projects. The purpose is to show how various methods are used together, and how they fit into the context of a complete project flow.

The examples shown in the book are focused on high-level concepts rather than code examples. There are several reasons for this. First, a great deal of the complexity of functional verification is in the interaction and complexity of various tools and methods. If the concepts of functional verification are understood, then the implementation is a more straightforward software development project. Second, there are a number of different languages and tools available. While the concepts behind the various tools are almost identical, the actual code styles differ. Finally, some of the methods are reasonably complex. Using actual code would result in large examples with a great deal of detail that would distract from the underlying concepts. By using more idealized examples, the focus can remain on the important issues around any particular technique.

As a result, the examples and detail level are chosen to provide a broad, high-level understanding of the strengths and weaknesses of various methods in functional verification, to show how various tools and methods may be used together, and how functional verification can fit into the flow of a complete project.

2

Definitions

Key Objectives

- Types of models

- Black box versus white box

- Definition of a test

Functional verification requires that several elements are in place.
It relies on the ability to simulate the design under test (DUT) with
a specific input stimulus, observing the results of that stimulus on the
design, and deciding if the results are correct. Figure 2-1 illustrates this
basic verification environment.

There are several components in Figure 2-1. The entire environment is
run in a simulator, since the device under test has generally not been
built yet. There are many different simulators and simulation languages
available. Two of the most common languages are Verilog and VHDL.
Both of these languages have a variety of commercial simulators
available. The basic purpose of the simulation language is to allow the
behavior of a design to be concisely described and efficiently simulated.

Abstraction Levels

The level of detail in the DUT description may vary based on the
abstraction level that is used. A more abstract model will describe the
overall behavior of the design, but may leave some internal details out.

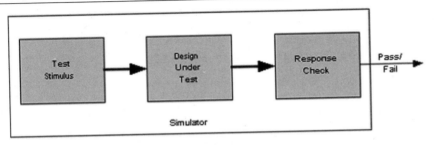

Figure 2-1: Simulation Environment

A less abstract model will provide more detail, and be closer to the actual implementation when the device is built. Usually, the more abstract models are faster to write and simulate, while a more detailed model is used to synthesize the actual device. It is not uncommon in complex designs to start with a more abstract model that focuses on the algorithmic and architectural issues of the system, then create a more detailed model that contains the information necessary to synthesize the actual device.

Behavioral Model

The highest abstraction level is usually referred to as a behavioral model. At this level, there may not be any timing information in the model. Some functions, such as error handling, may be missing. The goal at this level is usually to examine the basic operation of the design, and possibly the interactions among various components within the system. It is important that most of the tests and response checks are able to run with the behavioral model, since this permits both the model and the tests to be verified. As an example of a behavioral model, consider an integer multiplier. A real design may consist of a Wallace tree with multiple adders instantiated. A behavioral model can abstract all of that away. Listing 2-1 shows a behavioral multiplier in a pseudo language.

At the behavioral level, the abstraction may be very significant. The example in Listing 2-1 shows a multiplier that provides no

Listing 2-1: Behavioral Model

```
1    input    [7:0]  a;
2    input    [7:0]  b;
3    output  [15:0]  c;
4
5  begin
6    c = a * b;
7  end
```

implementation detail at all. In this example, there is no timing
information to indicate how many cycles it takes to perform a
multiplication, nor any pipelining information. The algorithm is
provided by the software language. There are many cases where the
algorithm of an abstract block may be provided by existing software
tools, even when another language is linked to the simulator to
provide the algorithms.

An abstract model may provide cycle-level timing information. In the
case of the multiplier example, a delay could be added to correspond
to the estimated cycle delay of the model. In cases where a cycle-
accurate model is required, then care must be taken that the cycle
delay built into the behavioral model exactly matches the number of
cycles needed for the implementation.

Register-Transfer-Level Model

A more detailed abstraction level is referred to as the register transfer
level (RTL). In many systems, this is the level at which most design
and verification effort is done. As the name implies, the RTL specifies
where the storage elements are placed, providing accurate cycle-level
timing information, but not subcycle timing, such as propagation
delays. This level still shares many constructs with software, but
requires several hardware constructs as well to simulate the parallel
nature of hardware. As a result, hardware description languages such
as Verilog and VHDL are commonly used for RTL modeling. At this
level, all of the functional detail is usually available in the model.

This is because the RTL is often the bottom-most layer for creating a design.

At the RTL level, the placement of the registers is specified, but the asynchronous logic may still be abstract. Figure 2-2 shows an example of an RTL diagram.

The hardware description language (HDL) version of this code will still have software constructs. Listing 2-2 shows a segment of RTL code to implement the counter above. As always, the code is idealized. In this case, the reset conditions have been left out.

Note that while the registers have been defined for this logic, the model is still abstract, since the incrementer and the control logic are still expressed as software constructs. Nonetheless, with all the registers in place, this model will be a cycle-accurate model of the actual

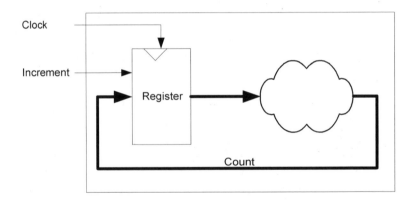

Figure 2-2: RTL Counter Diagram

Listing 2-2: RTL Counter Example

```
1   reg  [7:0]  count;
2
3   always @(posedge clock) begin
4      if (increment == 1) count = count + 1;
5      else                count = count;
6   end
```

device. From a simulation viewpoint, this model should behave almost exactly the same as the actual device will. This can be important when exact cycle behavior needs to be verified.

Gate-Level Model

Gate-level models tend to be the lowest abstraction layer used in the front-end design of a system. At the gate level, each individual logic element is specified, along with all the interconnections between the elements. For any reasonably complex design, this abstraction level is difficult to work with, since there is so much detail. Simulators tend to be quite slow at this level as well for the same reason. Despite the detail, there is still significant abstraction at this level, since the individual transistors are not described.

While most functional verification tends to be done at higher abstraction layers, there are times when a gate-level model may be important for verification. This will happen when the abstraction to RTL may hide some gate-level behavior that needs to be verified. This may occur in places where subcycle timing information becomes important, or where the RTL simulation may not be accurate. Some common examples where gate-level simulation may be of interest are for clock boundaries, where propagation delay, setup and hold times are important, or for reset conditions where "X" or "Z" logic states corrupt the behavior of the RTL abstraction.

Since these specific issues are usually a tiny fraction of the overall design, most verification is done at a higher abstraction layer. This not only improves simulation performance, but permits easier debug, since the more abstract models tend to be easier to read, understand, and debug.

Verification of a Design

In order to test the design model, the test must drive stimulus into the design under test, and examine the results of that stimulus.

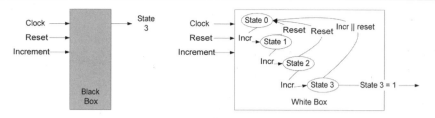

Figure 2-3: Black Box vs. White Box

The stimulus and checking can be done at the boundaries of a design, or it can examine the internals of design. This is referred to as black box versus white box testing.

To illustrate the difference between these approaches, consider a simple state machine that can increment or reset, as shown in Figure 2-3.

White Box Model

In a white box model, the test is aware of the inner workings of the device under test. This allows the test to directly monitor internal states to determine if the device is behaving correctly. For example, in Figure 2-3, the test may monitor the current state of the state machine. This allows the test to directly check that the state increments and resets correctly. The test obtains the immediate feedback about the state of the design when new stimulus is introduced, so errors may be reported earlier, and the test does not need to infer how the device under test is operating.

Even when a test is connected to only the periphery of a block, it may still be a white box test if it is operating with knowledge of the internals. If a test is written to cause a specific condition inside the block that is not specified in the design specification, then it is probably a white box test. As an example, a design may use a first-in, first-out queue (FIFO) to implement a function. If the test writer knows that the FIFO is four stages deep and writes a test for that, then the test is using knowledge of the implementation, rather than knowledge of the specification, and is thus a white box test.

Black Box Model

In the black box case, the test is limited to examining the input and output signals, and determining if the model is working correctly based only on information gathered from the outputs. In a black box model, the test environment either cannot or does not access the internals of the device under test. In some cases, the model may be encrypted, or otherwise be inaccessible. In other cases, the tests are written to look only at the inputs and outputs even when the internals are accessible.

The black-box-based test must infer internal operations without being directly able to observe them. In the example shown in Figure 2-3, the state can only be determined by asserting the increment signal until the State 3 output is asserted. There is no other measurement available.

As a result, a test may require more simulation cycles, and more logic to determine if the block is operating correctly. This is an additional burden on a black box test, but there are also some advantages.

A white box test is dependent on the internal structure of the logic. Any changes to the device may affect the test. One obvious example is synthesis. When an RTL model is synthesized to gates, the signals may change, preventing a white box model from running until it has been modified. Similarly, a white box test tends not to run on a design in multiple abstraction layers, since the internal workings of different abstraction layers may be significantly different.

A black box test, on the other hand, may run on a block that is designed as a behavioral, RTL, or gate-level model without modifications. Because it is looking only at the inputs and outputs of a component or design, it tends to be more re-usable as the scope of the design changes. This level of re-use may offset the additional complexity needed to infer the operation of the device.

There are other reasons for using black box models as well. First, a test is often focused on verifying the intent of a design, rather than the implementation. In this case, the details of how a design was constructed are irrelevant from a test viewpoint. The test is measuring only the final results, which would be found on the outputs. For this type of test, there is no reason to examine the internal state of the module.

A second reason that black box models are used is that they can separate the design from the verification. If a test is written by examining the inner workings of a design, it is possible that the test writer will write a test to verify a block based on reading the code from that block, rather than basing the test on a specification. This effectively corrupts the test so that little is gained from that test. For white box tests to be effective, they may use internal information, but care must be taken to ensure that the test is based on the specification, rather than on the RTL.

Both black box and white box tests have value. It is important to consider the goals and re-use requirements of the test to determine which is best suited for any particular situation. It is not uncommon for tests that are focused on the details of small blocks of logic to be white box tests, while those that look at broader multi-block issues are black box tests.

In some cases, a black box test will examine a few key points in a design, but for the most part use only the inputs and outputs. This is sometimes referred to as a gray box model, since it follows the basic concepts of the black box model, but with a few added outputs.

Definition of a Test

The simulation environment supports not only the device under test, but also the test stimulus generation and checking. The test itself may also reside in the simulator, written in the language that the simulator

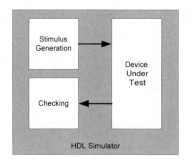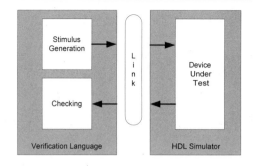

Figure 2-4: Tests Written in Simulator or in Verification Language

supports, or it may be written in a different language which is linked to the simulator.

A test that is written directly in the simulation language is easiest to understand since there is no additional software linkage. On the other hand, there are many verification constructs that may be useful when writing a test that are not supported by an HDL simulation language. Figure 2-4 illustrates what these two might look like.

There are several commercial companies that provide both an HDL simulator and a connected verification language, or a verification language that can connect to a simulator so that users do not need to deal with the link between the two languages.

Whichever language choice is used, a test is expected to drive stimulus into a device under test, and to observe the results in some fashion to determine that the DUT behaved as expected.

3

Methods for Determining the Validity of a Model

Key Objectives

■ Stimulus generation methods

■ Results analysis methods

A functional design must react correctly to a given stimulus. The process of validating the functionality of that design has two major components: driving stimulus to create a specific set of conditions inside the design, and measuring the reaction of the design to that particular condition to determine if the model behaved correctly. The effects of the stimulus on the design need to be understood, and the generated stimulus must be appropriate for the condition of interest.

The measurement of the reaction may be narrowly focused on a particular subset of the design that was intended to react, or it may be broad, checking how the entire design operated during the event.

Overview

Stimulus Generation Methods

There are different methods of generating stimulus that will run against a model of the implementation. Many of these methods have evolved from some old tools, and some make use of more advanced tools. Because methods vary widely in terms of how they perform tests, it can sometimes be useful to use more than one method on a single project where they are able to focus on different issues.

The majority of stimulus generation methods can be divided into a few basic categories. These are:

- Vector generation
- Stimulus capture
- Transactions
- Assertions

Results Analysis Methods

A test is only useful if the results of the test are always examined and analyzed. Simply driving stimulus will not provide any assurance that the design or the architecture are in any way correct.

The majority of results checking methods also fall into one of several categories:

- *Eyeballing* – Examining the results manually.
- *Golden files* – Comparing outputs to those stored in a file.
- *Assertions* – Making statements about expected behaviors.
- *Monitors* – Tracking outputs.
- *Predictors* – Comparing outputs to a model that predicts results.

Stimulus Generation Methods

Test Vector Generation

One of the earliest and simplest methods for generating tests is to simply specify exactly what values are to be sent to every input of a

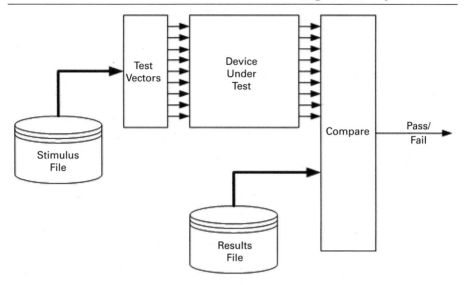

Figure 3-1: Vector-based Test

design. For synchronous designs, that would mean specifying the value of all inputs on each clock cycle. For asynchronous designs, it means specifying the exact times that each input will change.

The method for specifying the vectors is quite straightforward. For simulation environments, the vectors may be defined in a file that is read in as part of the simulation and is driven to all the pins of the device, as shown in Figure 3-1. Waveform drawing programs can provide some input simplification, and may ease the task of editing the waveforms since they are able to provide an easier format to view, save, and edit a waveform file.

The primary advantage of this type of technique is that it is simple to understand and simple to use. The exact stimulus is specified by the vector files, so there are no intermediate steps between what is seen in the waveform file and what occurs in the simulation. There are few conceptual hurdles to overcome in this type of environment. It is necessary only to understand the functionality of the design, and not much else.

Despite the ease of using this type of method, it is rarely used for anything but trivial designs. Many modern designs are quite large, in the range of millions of gates. With a large state space, which is defined by the number of registers in the design, and significant amounts of memory, the number of cycles required to simulate such a device is very large, often in the millions. Generating vectors of anywhere near that size is not practical. Once the vectors have been created, any modifications could have a significant impact on the vectors. For example, if the cycle timing of an interface were to change by a few cycles, then the vector files would need to be updated to reflect that change. Updating vector files could require significant time and effort.

To ease the vector creation task, there are some waveform tools available that allow the vector files to be specified as waveforms. By viewing the vectors as waveforms, it may be easier to match vectors to bus interfaces.

An example of a vector input is shown in Figure 3-2 as a waveform. This is a waveform of a write operation on a simplified processor bus.

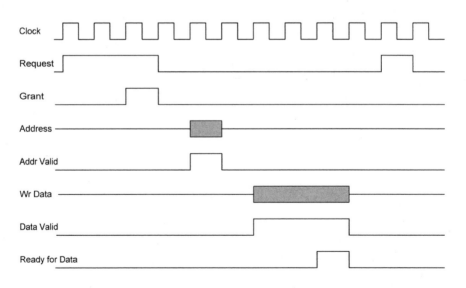

Figure 3-2: Simplified Processor Write Waveform

The stimulus necessary for a single write operation requires enough data for this to be a tedious job if done manually.

It would be possible to write a script that converts read and write operations into vectors or waveforms. This would be a simple form of transactions.

Stimulus Capture

In some cases, a device already exists that is similar to the design being created. In such a case, one can capture vectors to provide stimulus to an existing design. This technique is similar to the vector capture mentioned above, but a large number of vectors can be easily captured. Figure 3-3 shows how vectors may be captured by monitoring the inputs and outputs of a device that is running in a real system.

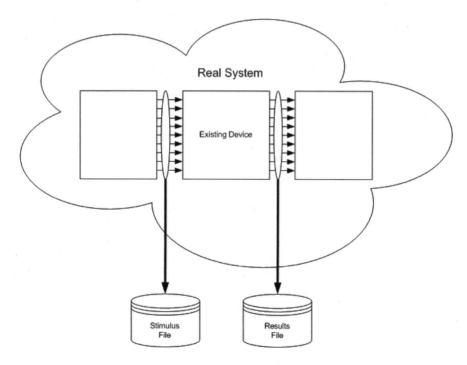

Figure 3-3: Stimulus Capture

In many cases, stimulus capture is slightly more complicated due to tri-state signals that are both inputs and outputs. Even with that complication, if an existing device exists, then stimulus capture provides a simple and effective means to compare the operation of the new design against the existing device.

Transactions

Transaction-based verification is a method used to provide a level of abstraction between the actual pins of the model and the test that is exercising those pins. In most designs, there are pins that can be organized into a well-defined group, where the relationships between those pins are understood. This allows a test to be written in a more abstract fashion, specifying a transaction to be done, rather than needing to specify all the details about how an operation is to be executed.

The simplest example is a bus, such as shown in Figure 3-4. In this case, the address, data, and control lines are all part of that bus, and the relationships among these signals are defined in a bus specification. A transactor can be used that will control all of these signals as a single group. Figure 3-4 shows the pieces of a processor bus transactor, and how they interact with the test, and the device under test.

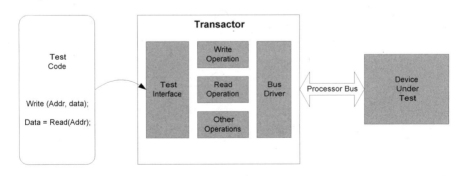

Figure 3-4: Processor Bus Transactor

26

The transactor provides an interface between the test code, and the processor interface in the device under test. The transactor itself contains a significant amount of code. First, it must implement the processor bus protocol. This may be done in Verilog tasks, or it could be done in a variety of other ways, depending on the tools and methods available. In any case, the knowledge of the processor bus protocol is built into the transactor.

The other critical part of the transactor is an application interface that allows another task or program to control the processor bus through a set of standard calls. This permits the test, or some other intermediate code, to perform all legal processor bus operations, read the results, check for status and error conditions, and potentially drive error conditions if that is required.

The test code in this example may be very straightforward code, perhaps code that performs a sequence of writes and reads, it may be arbitrarily complex, or it could be an intermediate layer, perhaps an interface to an instruction set simulator. Notice that the language and complexity of the test code is not important to the concept of a transactor.

There are several aspects of transactors that allow them to add structure to a functional verification environment. These include:

- *Encapsulation* – The transactor contains the knowledge of the protocol of the group of signals that make up the bus. All interactions with the bus are undertaken by the transactor. Any blocks that connect to the transactor, such as the test code in Figure 3-4, do not need any knowledge about the bus protocols.

- *Abstraction* – Since the test does not need to be involved with the bus protocol, it can be written at a higher abstraction level. In the simple case of the processor bus transactor, the test is able to specify operations in terms of reads and writes. It is not involved in any lower-level issues having to do with the bus.

- *User interface* – The transactor is able to provide a standard set of routines that a test can call. In the example of Figure 3-4, these routines are a read and a write call that are available to the test. This can be used to provide a standard application programming interface (API) that any number of tests can use, and that various transactors can share.

- *Modularity* – With transactors, the verification environment can be built from a set of parts. The transactor is the first example of a modular component. As an example, if the processor bus on the device under test in Figure 3-2 were changed, then the processor bus transactor would be swapped out for a different transactor, and the test should not be affected. There are some exceptions to this, such as tests that verify low-level bus protocols or error-handling capabilities, but modularity will generally hold for a large percentage of the tests.

With the large and growing cost of functional verification in many of today's projects, re-use of verification code is becoming critical. These four features provide the basis for achieving significant re-use within functional verification. This will be examined in more detail in Chapter 5.

Assertions

Assertions provide a different approach to testing a model. An assertion is a statement about a design that is expected to be true. The statement may consist of both logical and temporal components. The logical component is usually expressed as an equation, while the temporal component specifies under what time constraints the equation is expected to hold true.

While the exact format and abilities of the assertions are tool dependent, they are generally simple statements that define a rule between inputs, current state, and outputs. The rule may be both algorithmic and temporal in nature.

To provide an example, there is a simple state machine shown in Figure 3-5 with a few input and output signals. This represents a synchronous design with a single clock. While this example is contrived, it shows how it may take several clock cycles for the state machine to be reset, since it must travel through several states to reach the idle state.

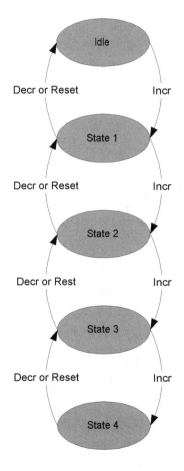

Figure 3-5: Simple State Diagram

Assertion statements are generally written in an assertion language, such as SUGAR, which provides a set of language constructs for assertions. A set of assertions for the diagram above could look like this:

1. When reset is asserted, state must be idle within four cycles.

2. When reset is asserted, state must be idle within two cycles.

If one assumes that only these five states are possible, then the first statement is true. From any state, if reset is asserted for four cycles, the idle state will be reached. The second statement is false. From State 3, idle would need to be asserted for three cycles to reach the idle state.

Notice that this example is somewhat contrived, since the assertion statement appears to be wrong if one assumes that the state diagram is correct. It is usually the other way around. It is also worth noting that assertion statements usually require the use of a white box method. In this example, the assertion statements are verifying a state machine that would usually be implemented inside a block of logic. A white box is needed in order to access the internals of the block.

Generally, assertions are statements about a small part of a design, and they are used to make algorithmic or temporal statements about signals or relationships between signals. A violation of an assertion statement is flagged, since this represents an error in the design.

Consider the processor bus diagram shown in Figure 3-2. There are many assertions that can be made about the relationships between various signals, for instance:

- There is a one-to-one relationship between request and grant. Grant may only be asserted after request is asserted, and similarly, if request is asserted, then grant must also be asserted after a finite number of cycles.

- There is another one-to-one relationship between address valid and grant, similar to the one between request and grant. Notice that there are generally at least two relationships between most signals. In the case of request and grant, one may be more obvious than the other. Not only is grant only permitted to happen after request is asserted, but there is also a second relationship, which is that subsequent requests may only happen after grant is asserted.

A sample set of assertions for the example processor bus waveform is shown below. The syntax of assertion statements is very tool dependent. For the sake of clarity, the examples here are English oriented.

- *Request without grant* – Grant must be asserted within *n* cycles of a request.

- *Address valid after grant* – Address valid must be asserted within *n* cycles after grant.

- *Address and address valid* – Address must have a non-Z value when address valid is asserted.

One can make a set of assertion statements that would ensure that the bus protocol is being followed correctly. With a complete set of assertions, any deviation from the defined bus protocol would be detected and reported by one or more assertion statements.

Assertions can also be combined to provide a more transaction-oriented view. For example, in the processor write operation shown in Figure 3-3, there are many individual assertions that can be grouped together to capture the entire write operation. Listing 3-1 shows an example of assertions that cover the write operation.

By combining assertions, the group of assertions, sometimes referred to as a regular expression, are able to verify the protocol of the bus operation, and to examine transaction integrity. In some tools, the assertions can be used to capture and examine transaction data as well as to monitor the protocol of each transaction for violations.

Listing 3-1: An Example Assertion Sequence

```
Write_operation: {
1: Request must stay asserted until grant is asserted
2: Grant must be asserted within <n> cycles of Request asserted
3: Address must be asserted when condition 2 is True
...
}
```

As more complex sequences of assertions are created, the advantages of using an assertion language become greater. A good assertion language is constructed to provide an efficient means to specify temporal, algorithmic, and sequential properties.

Static vs. Dynamic Assertions

Assertions can be validated either dynamically or statically. Dynamic assertions are tested during the course of a simulation run, while static assertions, which are sometimes referred to as formal assertions, use a separate tool to validate them. Some tools use an assertion language that can be validated statically and dynamically.

Dynamic Assertions

Dynamic assertion statements require a simulation-based environment. The assertion relies on stimulus to be generated and run in the simulation. If a pattern is detected that violates the assertion statement, then an error is reported. The dynamic assertion methodology may be implemented in a specific assertion language to provide a consistent temporal syntax. Alternatively, one could implement dynamic assertions in the same language as the design itself, and simply report any unexpected conditions.

For dynamic assertions to be effective, it is critical that the stimulus generation create the necessary conditions to check the logic. In the example in Figure 3-5, there were two assertion statements:

1. When reset is asserted, state must be idle within four cycles.

2. When reset is asserted, state must be idle within two cycles.

For the second assertion statement to be proven false using dynamic assertions, the stimulus logic would need to put the state machine into State 4 and then set the reset condition true for multiple cycles to prove the second assertion statement. If the assertion did not trigger, it may mean that the logic is valid, or it may mean that there was insufficient stimulus generation.

To help answer that question, one can provide some information to determine if there was sufficient stimulus. In the case of the second assertion, it is possible to detect if a reset condition was driven for two cycles after the state machine was in State 4. This ensures that the assertion statement verified that at least one path worked successfully. It does not guarantee that all paths will meet the assertion requirements.

Static Assertions

Static assertions are not simulation based. Rather, they use a mathematical model of the design to determine if there are any possible scenarios where an assertion statement could be violated. If such a scenario is found, most tools will provide an example scenario that illustrates the violation. Because formal methods are complex, a commercial tool is generally required if static assertions are to be used.

When a static tool is used on the example in Figure 3-5, it will analyze the design, and find any conditions where the assertion statements are false. The static tool will examine the statements such as the two listed above, and report that statement 1 is not violated, while statement 2 could be violated. For these statements, it is the static tool that will find an example to show where the second assertion statement fails, when the starting state is State 3 or State 4.

Unlike the previous methods, assertions provide their own results analysis. Part of the assertion specifies what the desired outcome is, so there is only one step, which is to define the assertion from a protocol viewpoint.

Advantages of Static Assertions

There are several major advantages to static assertion statements. The static methods allow the assertions to be checked against the entire state space. That is, the assertion will be verified against any possible state that the design could be in.

As a result, there is no need to create and drive stimulus into the device under test in order to create the conditions of interest. Since stimulus generation can be a very time-consuming task, this can be a significant advantage. For example, in Figure 3-5 it may be required that all state machines be in the idle state in two cycles. To catch this problem with simulation techniques, it would be necessary to create the stimulus to put the state machine in State 3, and then assert reset for two cycles. While this can certainly be done, it is quite conceivable that such a test is missed if an engineer did not realize that this was a possible failure scenario. Similarly, the verification engineer may have no means of even knowing that a particular scenario was needed if it was not in the design specifications.

Finally, static assertions provide a one-step process for functional verification. There is no need to create stimulus. This method bypasses that time-consuming step.

Disadvantages of Static Assertions

There are significant disadvantages with static assertions as well. This method focuses on the behavior of specific blocks and protocols. The assertions are about specific behaviors of a particular block, without regard to how the block interacts with other blocks. One can make broader assertions about multiple blocks, but they often become much more complex, and often require too much computing power and time to be an effective tool.

Systems tend to be structured and designed as a sequence of blocks that interact with each other. The assertion statements do not often fit naturally into this type of flow. As a result, it can be difficult to correlate the architecture of the system to a set of assertion statements. Simulation-based techniques are often more closely aligned to the way that systems are structured and designed. This may make it easier for simulation methods to recreate the way that a system will be used in a real environment.

Static assertion methods are quite different from simulation-based techniques, with different strengths and weaknesses. Because of this, a project may be well served by using both methods together, exploiting the strengths of each.

Where static and dynamic assertions provide a powerful low-level check of specific design components, and bus protocol compliance, simulation methods are more closely aligned with the way that the architecture intends for data to pass through the system. This can provide better insight into how the design will behave when it is implemented and running in the actual environment.

Results Analysis Methods

To validate that a design works as intended, it is important not only to create a particular condition through stimulus generation, but also to measure that the design functioned properly in the scenario. This requires examining the state of the design to see how it reacted. Examining a reaction could be very narrowly focused to see if a specific action occurred, or it could be very broad, looking at how multiple parts of the design acted.

Determining the correct operation of a design can be complex. Even when the basic operations and interactions are well understood, there is an enormous amount of information contained in a simulation.

Manual Results Analysis

There are many methods that are used to capture and analyze the outputs of a simulation. Only some of them are likely to be effective in larger designs. Results analysis can be done manually, or through a variety of self-checking tests.

Eyeballing

Eyeballing is a term used for the manual examination of a simulation. This can be done using waveforms or text output. In either case, whether a test passed or failed is determined by examining the output manually for correct behavior.

The primary advantage of this method is that it can be easy to perform the first time. The output can be examined interactively to see if the device behaved as expected. Any operations that are not understood can be explored on the spot to determine if they are correct or not. No time is spent predetermining the outputs, so it is possibly the fastest way to get a single pass/fail determination on a small design.

Despite the simplicity of this method, it tends to be unacceptable for any project greater than a few hundred gates for a variety of reasons. The first is that human examination is not a repeatable process. While the first examination may be quite thorough, the second, third, or twentieth will each be successively weaker, simply due to boredom and familiarity. What that means is if a new error is introduced into the design, there is a significant chance that it won't be detected.

The second major issue with this method is that it is difficult for humans to analyze the information that is contained in a simulation. Even on a simple bus, there are a large number of signal interactions and dependencies that can be very difficult to examine by hand, although abstraction can help a little bit in this regard. Examining each signal by hand is an almost impossible task. By storing or post-processing the outputs so that one looks at transactions rather than signals, the task of eyeballing the results is at least somewhat easier.

Finally, there is no ability to perform an automatic regression test. A regression test is a run of all existing tests to determine if the design works at least as well as it did in the past. This is useful as design changes or additions are being made, particularly when several people are adding to the design. The regression test ensures that no new errors have been introduced into working code, and that multiple sets of changes do not conflict with each other.

For these reasons, eyeballing methods are unlikely to provide satisfactory results for functional verification.

Golden Files

This is a method of comparing the output of a simulation to a stored version of the expected output. The stored output is usually a file, which may contain cycle-by-cycle information on every pin, or possibly a higher-level storage of every transaction.

The first issue is how the golden file is created. In the simplest case, it is created from the output of a previous simulation run. This avoids some of the disadvantages of eyeballing. As long as the design does not change, then a comparison with previous simulation results will work quite well.

Another way to create the golden file is by specifying the contents beforehand. This requires that the outputs of the model are predicted beforehand, and then compared to the simulation run. The major disadvantage of this method is that all of the simulation output needs to be thought about beforehand. For any complex design, this is impractical, since there will be thousands to millions of output cycles.

However the golden file is created, there are still multiple critical disadvantages with this method. If a design changes in any significant way, such that the output is shifted or modified even slightly, then the

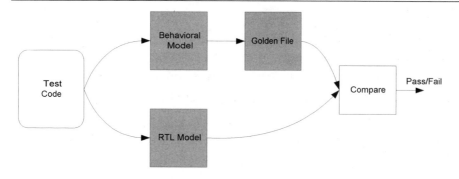

Figure 3-6: Comparison of Behavioral and RTL models

comparison will fail. This type of modification tends to occur quite frequently in the course of a design as registers need to be added or removed. Each time a small modification causes a change in the output, a new golden file is needed. This brings up the problems previously mentioned with determining the validity of a new golden file.

There are some applications where golden files may make more sense. Where there are two models of a design, then golden files may be used to determine if the models are functionally equivalent. For example, if there is a high-level behavioral model and an RTL model, golden files are a way of comparing the outputs between these two models without having to run them concurrently, as shown in Figure 3-6. Note that this does not help with determining whether the operation of both of the models is correct, only in determining that the operation is the same between the two models.

Automated Results Analysis

Where assertions differ from the previously described methods is that they are self-checking. An assertion will continue to check for an invalid condition as more tests are added. If there are changes made to the design, as long as the protocol definition does not change, the assertions remain valid and continue to detect any violations.

In any reasonably complex project, it is common to have near-continuous changes in tests, in the design RTL, and sometimes in the architectural specifications. The ability of tests to be self-checking is valuable, since this allows them to be resilient in the face of a constantly changing environment. Since a considerable effort goes into any type of results analysis, self-checking assertions can relieve the need for excessive project resources to manage and update changes. Golden files are often said to create self-checking code, since they provide an expected output that can be compared to the actual output, as shown in Figure 3-1. However, when the golden file itself is checked manually, then the point is debatable at best.

Assertions are most likely to be useful in verifying that specific sequences occur at a particular point in a design. That may be on a bus, or in a state machine. While this is clearly valuable, a design is usually intended to pass data through a system in one fashion or another. Assertions are often not as well suited to follow data as it passes through a system as simulation methods might be. In simple cases, where data is sent into one side of a system and received on the other, assertions can be written that may follow the data. More complex systems do not always follow such simple rules. Exception handling, interrupts, error conditions, or queuing all tend to cause a linear sequence of events to become non-linear. Defining a complete set of assertions for a more complex sequence is usually not practical for a complex device.

This does not mean that assertions are not valuable. It may mean that assertions alone are not the ideal way to verify a system.

Another issue with assertions is the sheer number of them required to perform any level of complex checking. Even with the simplified processor waveforms shown in Figure 3-2, a significant number of assertion statements would be necessary to adequately verify the bus protocol. Simply having a large list of assertion statements would quickly become unmanageable. Hierarchy and structure of assertions is necessary to organize the statements.

Protocol Monitors

Standard interfaces are often used in a design. An industry standard interface, such as a PCI bus, is one example. Or, a standard interface may be a proprietary bus that is defined within the scope of one or several projects.

Protocol monitors may be used to check that the bus protocols are being followed at all times. A protocol monitor may simply be a group of assertion statements that have been written to follow the protocol of a particular bus, or it may be code, written in the simulation language, that checks for valid sequences. By connecting this monitor to a particular bus instantiation, those statements will now be able to verify the integrity of that particular bus. In Figure 3-7, a design is shown with a variety of buses. Notice that multiple instantiations of a

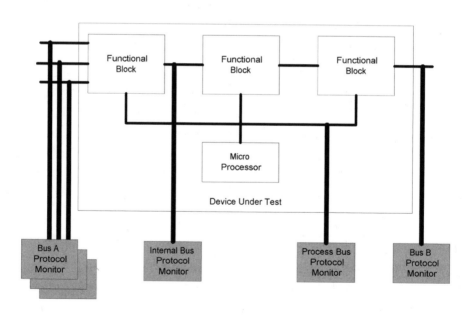

Figure 3-7: Protocol Monitors Connected to Interfaces

single bus may be monitored with multiple instantiations of a protocol monitor.

Protocol monitors may be used for a variety of functions. The first is to monitor a bus and report any protocol violations. Protocol monitors can also be used to collect statistics on a particular bus. By recording the types of operations that occurred, the frequency of operations, or status issues such as occurrences of parity errors or timeout conditions, it is possible to collect and report bus usage information. In some cases, the actual operations are recorded to allow for tracking of operations through a system. This recording function can be particularly useful for measuring system performance. The implementation of a protocol monitor may include a collection of assertion statements for verifying the validity of transactions. A protocol monitor has several characteristics in common with transactors that allow for re-use in and between projects:

- *Encapsulation* – The knowledge of the bus protocol is contained within the monitor. It is possible to plug a monitor onto the bus without needing to have detailed knowledge of the bus protocol. More detailed knowledge is needed only if the monitor detects a protocol violation on the bus that must be fixed. The protocol knowledge is encapsulated in the monitor, so that a user does not need to be concerned with the details of the bus.

- *Abstraction* – Error-reporting mechanisms and statistics gathering do not need to understand or follow a particular protocol. With protocol monitors, any of these devices are able to rely on the monitors to provide the details needed. For example, on a PCI bus, the monitor may report the bus operations, such as read and write, and leave out the details of the protocol.

- *Application interface* – A protocol monitor will generally have two interfaces. The first is to the bus itself. This is generally well defined by the bus protocol itself. The second is an interface for other routines to access the data from the monitor for error reporting or statistics gathering.

- *Modularity* – As with transactors, a protocol monitor can be instantiated anywhere in a design. It is a self-contained unit that can be used when and where it may be needed.

Protocol monitors are one of the simplest devices to re-use because they almost naturally meet the critical requirements for re-use. Even though they may be easy to use, it is still a significant effort to create, test, and debug a complex protocol monitor. The ability to re-use verification code can represent a large savings of the total engineering time in a project.

Predictions

Assertion statements provide one way of specifying the legal behavior of a block of code or a bus protocol. While this can be quite useful for a signal, or a small block of code, an assertion statement can be somewhat limiting in its ability to provide self-checking of a larger system.

In larger systems, multiple blocks tend to connect in a coherent fashion. There is a flow through a series of blocks with data transformations and control interactions, defined by an architectural structure.

From a system viewpoint, a test is used to validate that an operation is successful as it passes through a design, and that the blocks within the design interacted correctly with each other. This type of test is generally written as a sequential test, rather than a series of assertion statements.

Without the help of assertion statements to determine if the results of a test were successful, a different method must be used to determine the desired outcome of a test. There are several methods that may be used to predict the desired results.

Reference Model

For some projects, a reference model may be available. This can be in the form of a behavioral model, or a model from a previous project.

A reference model is often created from an architectural specification, and it will follow the data transformations and interactions of the system. As a result, when transactions are injected into a reference model, it should react the same as for the actual design. The advantage of a reference model is that all predictions of correct behavior are provided by this one model. It provides a very simple way to predict the correct behavior of a design.

- *Cycle-accurate reference model* – The reference model may be cycle accurate, which means that it will provide the same outputs on every cycle that the design is intended to provide. For a cycle-accurate reference model, the comparison function is straightforward. The comparison of the models is done on every cycle, and any difference in the output signals can be reported immediately. There may be some additional logic to deal with a few special cases, such as power-up sequences, but once synchronized, the comparisons are clear-cut. Figure 3-8 illustrates how a cycle-accurate reference model can be connected to determine if the output of a design is correct.

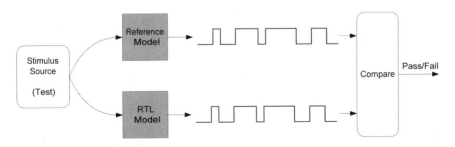

Figure 3-8: Cycle-accurate Reference Model

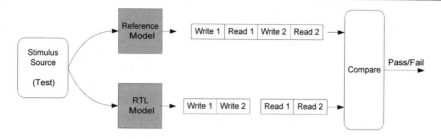

Figure 3-9: Architectural Reference Model

- *Behavioral reference model* – Alternatively, a more abstract reference model may only specify the outputs of the design to a transaction level. This would specify what the outputs are to be, but not exactly which cycle they are to occur on. In some cases, the order of the operations may not be quite the same either. Figure 3-9 illustrates a reference model that predicts the operations that will occur, but does not predict the time or ordering of the operations. A behavioral reference model may be driving signals, or it may simply specify when transactions would occur. How a model is written will have an impact on how the model is used to predict the output of the design. A highly abstract model may have more differences in output, and the comparison function must compensate for that.

In either case, to use a reference model successfully, it is important to understand the accuracy of the model, and how well the reference model will predict the design behavior. Things to consider are out-of-order transactions, delay difference through the models, and areas that are not implemented in the reference model.

There are several disadvantages to a cycle-accurate reference model. The first is the necessity of writing the model in the first place. This can be a time-consuming task, which may only make sense if the model is also needed for other purposes in the project.

Another disadvantage is that the accuracy of the model is of critical importance. A cycle-accurate model is wonderful if it works and is supported, since it provides a simple predictive model for the verification of the design.

A third drawback of a cycle-accurate model is that it requires a great deal of detailed knowledge to write it. Every cycle delay in the design must be known and taken into account in the reference model. In the event of even a minor register modification, the cycle-accurate reference model must be updated to reflect the change. An error will be detected any time there is any difference between these two models. For every difference, the source of the difference must be determined, the correct operation must be determined, and one of the models must be fixed. Not only is this a time-consuming task, and a drain on project resources, but it can be a source of frustration to engineers that must spend time on multiple models.

In contrast, a behavioral reference model is less accurate at a cycle level, yet usually far simpler to write and maintain. In many projects it is written only with the system architecture in mind, rather than using any information about the implementation of the design. Differences in the outputs between the behavioral model and the design depend on the approach used by the behavioral model. Because the behavioral model tends to be abstract, the outputs of the behavioral model and the design may look quite different.

Minor differences may include cycle timing differences, causing the reference model to drive transactions earlier or later in time. More substantial differences could include a different order of transactions, or some differences in the transactions themselves. This would be expected if there are algorithms that are similar, but not exactly synchronized, such as random number generators, or schedulers. Determining what and how to compare is an important part of using an architectural model for reference-based designs. Caches and queues frequently cause the order of operations to be altered, as shown in Figure 3-9.

Because of these types of differences, the comparisons of results may be more involved with a behavioral model. The comparison function must take these changes into account when determining if there is a match between the outputs. That additional effort is often balanced by the far simpler effort of creating and maintaining a highly behavioral reference model.

The critical issue for the reference model is that it should accurately represent the desired high-level behavior of the system, since it is being used to validate a design. One can argue that any difference in behavior between the design and the reference model will be detected during simulation. The risk with this is that an error in the reference model could cause the RTL design to be modified to match the incorrect behavior of the reference model if the nature of the error is not properly understood.

Self-checking Tests

Another method for checking results is to embed the information directly into the test. Since the stimulus is being defined by the test, usually to invoke certain behaviors in the design, the test also has knowledge about what the expected behavior is, and can therefore predict the results.

How difficult this is depends on the device that is being tested. Possibly the simplest example of predictive tests is those used for registers and memories. A test of a memory is shown in Listing 3-2. It can write many locations, and then later read them back. The test is able to predict the results of the read operations by remembering the write operations. Any miscomparison of data between the write and the read operation indicates an error. This is a very simple example of an architectural test. It is looking at architectural functions rather than

Listing 3-2: A Self-checking Memory Test

```
Memory_test_function

for  (i = 0; i < 20; i++) begin
    write (address = i; data = i);
end

for  (i = 0; i < 20; i++) begin
    j = read (address = i);
    if (i != j)  report_error;
end
```

any specific implementation issues. It is also a self-checking test since it verifies that the design implemented the defined architecture.

The memory test, while simplistic, is a good example of a test that is written against an architectural requirement. The scenario to be tested is defined in the test, and the expected output can also be defined in the test. With a reasonable level of abstraction from transactors, which supply the read and write functions, a self-checking test can be used even in larger complex environments.

There are several advantages to self-checking tests. A test can be written to check a particular set of features of the architecture. The test and the checking criteria are kept together.

Tests can be written from an architectural viewpoint. With the use of transactors, they can be written to be resilient to small changes in hardware, since what is checked can be architecturally oriented rather than implementation oriented.

It is often easier to measure performance-oriented issues through this type of test as well. Performance is generally measured in terms of transactions passing through a system. A test that is dealing with transactions into and out of a system is well suited to measuring the delays of the system and correlating those delays to architectural requirements to determine the pass/fail status of the system.

There are disadvantages to this approach as well. The primary one is that the more abstract tests require the support of transactors, and an

47

environment that provides the connections between the test, the transactors, and the system under test. This is a non-trivial amount of code that must be created or re-used from some other source.

Summary

There are a variety of methods for validating a functional design, some of which are effective in the realistic, complex projects. Assertions and simulation-based methods each have different strengths and weaknesses. The methods are not mutually exclusive. It can be effective to use a combination of approaches, depending on the size of a project, and the availability of tools and resources.

There is no single best method for generating stimulus or analyzing the results. In each case, it is important to understand the abilities of each of the methods, how stimulus is to be created, and the ability to capture and correlate the results with the architectural intent of the system.

Because the different approaches have such different strengths, the additional costs of a hybrid environment may be small compared to the productivity gain that multiple approaches can provide to a reasonably complex project.

4

Structure and Re-use in Functional Verification

Key Objectives

- Structural elements

- Buildup of complex environments

- Re-use of verification components

- Block- to system-level verification

The methods discussed in Chapter 3 can be used in realistic projects even though the examples were highly simplistic. In this chapter, structure will be introduced to show how the same methods are able to work with complex projects to maintain a manageable environment.

A structured environment is partitioned into a set of functions that allow the overall complexity to be broken into manageable parts. For the components to work well in the larger environment, the same features that were discussed in Chapter 3 are required: modularity, abstraction, encapsulation, and well-defined user interfaces.

These same features that provide a level of manageability in the verification environment also allow blocks to be re-used both within the project at different levels as well as between projects that have some overlapping architectural definitions.

Structured Elements

Simulation-based verification environments are almost always built upon a set of structured elements due to the power and flexibility that they are able to provide.

Advantages of Structured Elements

As with any complex software project, there are many advantages to using a structured approach. Among those advantages, but by no means the only ones are:

- Capacity to manage complex interactions

- Re-use of verification tests and blocks

- Ability to scale the project

- Ease of test writing

- Understanding of functionality

Transactor Structure

From a structural viewpoint, there are several key components of the transactor. Figure 4-1 shows a transactor as it might be connected to a device under test. Note that there are several key interfaces to the transactor: the bus that the transactor interfaces to, an interface for tests to access the transactor, and an internal interface that allows these two components to communicate with each other.

The top interface is the standard protocol. This may be a predefined industry standard interface, or a proprietary interface that is used within a particular product or family. In either case, it is generally a well-specified, pre-existing protocol.

At the bottom of the transactor in Figure 4-1 is a test interface that allows the tests to communicate on the standard protocol through the transactor while maintaining the abstraction and encapsulation.

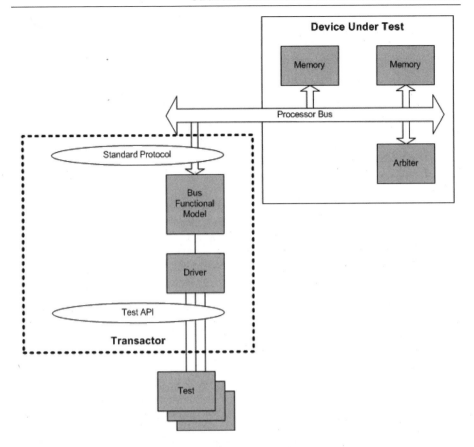

Figure 4-1: Transactor Interfaces

This interface can be made to be relatively generic, so that many transactors use exactly the same interface.

An internal protocol between the test interface and the bus functional model may also be used if the test language is different from the device language, or if there is an ability to share subcomponents between different transactors. This is likely if the tests are written in a verification or other higher-level language than the RTL uses. Some languages will attempt to shield users from this intermediate protocol. A clear transactor structure, with a well-defined test application programming interface (API), allows tests and transactors to be written in a similar fashion, and encourages future re-use between transactor designs.

Parallelism of Tests

Figure 4-1 shows multiple tests connected to a single transactor. The ability to have multiple tests run in parallel often has several advantages.

Combination of Tests Mimics Actual Traffic

Even when there is only a single transactor in the verification environment, there are often multiple streams of activity. A single processor may be expected to run an operating system that supports multiple processes. Each process will have its own independent set of instruction fetches and data operations that generate multiple streams of bus traffic.

If the processor itself has a cache, that cache will perform its own set of instruction and data prefetches and writebacks. Internal caches will affect the locality of references, and will alter the traffic patterns seen on the processor buses. Modeling traffic patterns on the processor buses may be easier with multiple tests running in parallel and sharing the transactor.

An example of how this can work is shown in Figure 4-2. Here, multiple tests are performing operations through a simple transactor interface. As a result, the operations will be sequential, but the bus operations will alternate between tests. This will result in bus traffic that is not simply incrementing through an address range. Parallel tests will often be orthogonal to each other so that the tests do not interact with each other. This could be done by using different memory ranges for each test, or by having each test focus on different components within the design so that the tests do not share any resources.

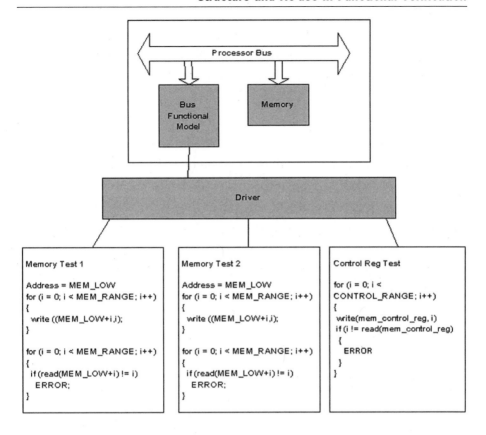

Figure 4-2: Multiple Tests Running in Parallel

Note that in this example, the bus functional model is inside the device under test. This may happen where the processor is a part of the device, which is frequently the case for system-on-a-chip (SoC) designs.

Modeling bus traffic may be important to understanding how well the caches operate. A cache depends on locality of memory addresses to be effective. Sequential addresses are not realistic tests for cache operations. Similarly, random addresses are also not representative of a real system. A random distribution of addresses will show that the cache is not effective, while a high degree of locality in a simulation will indicate that the caches are overly effective, and show the bus utilization as being low. By running multiple tests, it is often easier to mimic realistic instruction and data fetch streams that more accurately exercise the caches and buses.

Simplify Test Writing

While any sequence of operations can be created by a single test that generates any sequence of events, this can be a very complex test to write. When there are multiple sequences that need to be mixed together, it is often simpler to write a test for each individual sequence, and mix those individual tests together, as shown in Figure 4-2. The mixing may be simply interleaving the tests, or it can be an algorithm that determines which test is to run when. Each test can remain simple, and perform self-checking, yet the combination can produce a complex bus sequence.

Multiple Transactors

In all but the simplest system, there are likely to be multiple transactors. These may be multiple instantiations of a single transactor, or several different transactors, or a combination.

Transactors may be used for various system components. For example, a memory may be implemented as a transactor. This permits a sparse memory model to be used to efficiently model a large RAM by allocating storage only for those sections of the RAM that are actually used, rather than allocating storage for the entire RAM.

A memory transactor may also provide an easy way for tests to load or verify memory contents quickly at any time during a test. For systems with cache coherency implemented, direct access to memory can simplify the coherency test writing.

Multiple instantiations of the same transactor is also quite common. Figure 4-3 shows a dual processor system with cache coherency and

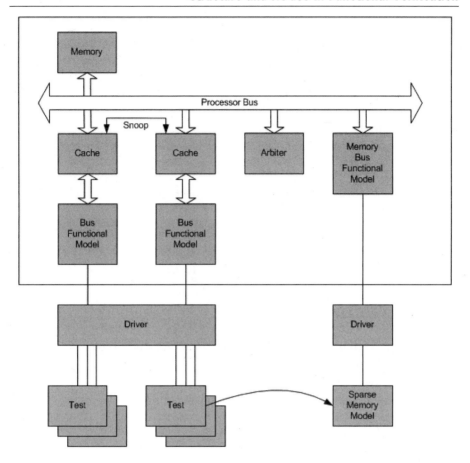

Figure 4-3: Example—Two Processors with Cache

a memory transactor. Again, the bus functional models are inside the device under test in this example, as would be expected in an SoC design.

In this environment, many simple tests can be used to create a highly complex test sequence. As an example, consider a cache coherency test that has multiple instances running on each processor. A coherency test works by forcing multiple processors to share a particular cache line. Within each cache line, one word may be used to indicate which instance of the test owns the line, and another word is used to store a signature to ensure that the line has not been modified. A simplified version of such a test is shown in Listing 4-1.

This simplified test attempts to lock a semaphore, and upon obtaining the lock, puts its unique identifier into a shared data word. It then releases the lock and repeats the process. This is a self-checking test that can force processor caches to share cache lines. This simplified test would catch some coherency issues, but it should be noted that a complete lack of coherency would not be detected in this test.

By having each processor execute this test, the caches are forced to snoop, invalidate, and fetch the lock line frequently. By using many cache lines, with each processor running an instance of the test for each cache line, the test will generate complex cache and bus

Listing 4-1: A Simplified Cache Coherency Test

```
// lock_word = first word of cache line
//             0 indicates unlocked

// test_and_set = atomic write operation. Only writes if the
//                lock_word is 0.  If unlocked, writes data.
//                returns the new data in the lock_word

forever {
  // Attempt to lock the cache line
  if (test_and_set(lock_word,my_id) == my_id)
  {
    // Cache line locked, so put my_id into the data word
    data_word = my_id;
    // wait some time, or do some other test or flush cache line...

    // verify that noone else has used this cache line
    if (read(data_word) != my_id)
    {
      ERROR;
    }

    // release cache line
    write(lock_word, 0);
  }
  else
  {
    // Another test has cache line
    wait some time;
  }
}
```

interactions. It is quite common for a simple test like this to generate massive traffic, and detect quite a few coherency issues.

Highly Parallel Environments

As the size of the device under test grows, there are often significantly more transactors needed to inject stimulus into the system. The increase in complexity may affect the determination of functional validity.

An example of a highly parallel system is shown in Figure 4-4. In this case, there are a number of processor transactors that reside on a split-transaction bus. This type of bus allows each processor to issue an operation and an address in one cycle, and release the bus until the data device is ready. At that point, a single data cycle is issued on a data bus to complete the processor transaction.

Because a split-transaction bus is not kept busy between the address and the data cycle, there may be multiple transactions in process at

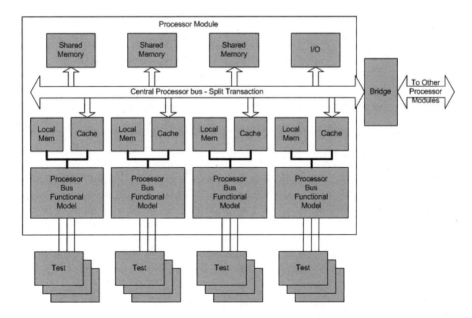

Figure 4-4: Highly Parallel Processor Environment

any given point in time. Many existing processors are able to issue four or more concurrent operations. This example shows a four-processor module that may be connected to other processor modules. A single processor module would be able to have 16 concurrent operations, and multiple modules would be able to support even more than that.

Running multiple instantiations of the memory test shown in Figure 4-2 and the coherency test shown in Listing 4-1, along with other tests, will create a complex stream of data on the central processor bus, all done with simple self-checking tests.

As before, functional verification will often be set up to model the types of traffic that might be seen in real systems. With the concept of parallel tests, each test may be quite simple to write and run. Many tests that are written for the example system shown in Figure 4-1 would be able to perform useful functions in this environment as well. Other tests would most likely be focused on the coherency and tagging aspects of this type of system.

The traffic that this type of environment will create may be quite complex. If there are just four tests, and a single processor module, and each test is issuing alternating reads and writes, Figure 4-5 shows what the processor bus traffic might look like.

Note that this traffic pattern is highly simplified. In a real test, there would be far more than four concurrent tests, and most of them would be issuing more complex sequences than writing and reading back data. Even the simple tests shown in previous examples will cause complex bus sequences. Combined with more complex tests, it can be difficult to determine the cause of the failure. It may be due to a design problem, an incorrectly written test, or an unexpected interaction with another test. Determining what operations happened when may ease the debugging task.

While any one test is still able to be self-contained and self-checking, each test is looking at a single component of the system. The

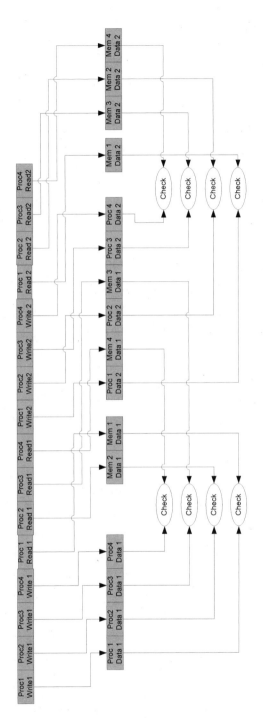

Figure 4-5: Tracking Bus Traffic

environment must be verified not only for correct operations, but also for any performance goals that are specified in the system architecture.

Highly parallel systems tend to have specific performance goals that may include throughput, latency, and fairness requirements. Any single test will be able to measure the performance of its operations, but that performance will be influenced by the number of tests and the types of operations the other tests were performing. Since there may not be a single point in the design where all operations may be measured, storing the transaction information may be useful.

Database of Transactions

Storing transaction information in a single location for later analysis is frequently referred to as scoreboarding. This may be a list or a database that contains information about all transactions that have been injected into the device under test. The term "database" is being used loosely in this case. Transaction information could be stored in a simple data structure or list that is searched. In some cases, a more classic database may be used in order to search for relations between transactions.

The goal of a database is to store all of the transactions that have been issued, along with all relevant information about that transaction, in a single central place. This will store not only the transaction as it is injected into the system under test, but also all of the responses. In the example of Listing 4-1, each operation is stored when the bus functional model drives it. Specific details of the transaction would probably include the time, which transactor, the address, and the operation. During the data component of the transaction, the database would store the time, the data, and which module responded, along with any status or error conditions.

It is possible to store all relevant information in a list of data structures for simple environments. What is important is the ability to store, search, and retrieve transaction information. As an example, if one just wants to find transactions that happened at certain times, or that had a specific address, then a list of transactions is likely to be quite effective. In more complex systems, different transactions will contain different sets of information, and linking between transactions may be important for debugging and tracking. In these cases, a more complex storage system may be worthwhile. A database is often more appropriate when analysis is to be performed on multiple pieces of data simultaneously. A database allows any data or cross-product information to be extracted easily.

Performance Checking

A significant amount of architectural and design effort is spent on achieving performance optimizations. Functional verification often must be able to analyze and confirm the performance goals.

There are a number of different performance metrics that may be specified in a system architecture, all of which need to be tested and measured in a verification environment. A database of transactions can make the process of verifying performance metrics easier. This may happen when performance measurements are made across a distributed design. Using a database of transactions as a central point that tracks when all transactions enter and exit various points in a system can simplify the task of performance analysis. It may not always be clear why results are being achieved, or even how to measure overall performance. With a database of transactions available, it is possible to explore how the system performed, and where improvements could be made.

There are a variety of performance metrics that often need to be examined. These may include the more direct throughput measurements, such as the number of transactions executed per second, or the utilization of any particular resource, such as a bus. Some systems will require more implementation-dependent measurements such as ensuring fairness across processors, or that the latency of operations does not depend on the memory unit being accessed.

These types of verification checks are often effective in finding logic errors in modules such as an arbitration module, where the individual operations are all legal, but a slight bias shows up when the transaction performance is compared between processors or memories.

The types of operations that can be measured from the example shown in Figure 4-5 include:

- Bus utilization

- Transaction latency to local module memories per processor

- Transaction latency to remote module memories per processor

- Transaction latency to memories from local processors per memory

- Transaction latency to memories from remote processors per memory

- Cache writeback latency to local memory per cache

- Cache writeback latency to remote memory per cache

Depending on the system requirements, the average or maximum times may be of interest. These types of calculations are simple to make using a database of transactions, and can provide an early indication of subtle problems that could affect system performance, but may not be caught by more traditional monitoring functions.

Database of Transactions Can Aid as a Debugging Tool

A database of transactions can also be valuable when a test indicates that an error occurred. In a complex environment, there may be hundreds of transactions that occur between the time that a condition is set, and the result is checked. Having a single location where all transactions are stored may be helpful in finding which transaction may have caused the condition.

As an example, consider a test that writes a location in memory. At a later point, it will read that location back to verify that the memory is still valid. If the read operation returns different data, an error is flagged.

In this example, the cause of the error may be due to a design failure, or it may be due to an unexpected interaction with another test that is running. A database of all transactions provides an easy way to look for potential causes of the memory error.

- *What tests wrote to that location in memory, and when?* By finding the sequence of transactions that accessed the memory location, any test error that caused the memory location to be overwritten would be highlighted. This would show a history of all accesses to that particular memory location.

- *What writes occurred with the faulty data?* If a history of memory references didn't find the problem, then one could explore where the faulty data came from. One possibility would be to examine all transactions that contained the incorrect data. If a write to another memory location is found with the faulty data, it is possible that there is a memory addressing problem that caused the data to be stored in the wrong location.

Even in this trivial example, a single repository of transactions may be useful. In more complex transaction sequences, finding the source of a problem may involve searching through thousands of transactions. The ability to sort and query the database of transactions can be a powerful debug tool.

Determination of Pass/Fail Status

Finally, in some situations, a single test may not be able to determine if the design under test passed or failed. A more global view of the system may be required. Under these situations, a test may be responsible for injecting transactions, but determining the correct operation of the system may be made after the tests have completed by examining the database of transactions. The most common example of this is in systems where transactions do not always complete. Error conditions may cause transactions to cancel, or heavy traffic may cause transactions to be dropped.

In this type of scenario, if a test were to simply check the results, it would report a failure, even though the logic functioned correctly. The database would be used to correlate the test results with the error condition, to determine that a particular transaction should have failed.

Data Traffic Domains

To this point, all of the examples of transactions have revolved around processor operations. They provide a simple structure, often consisting of only three fields: address, operation, and data. They also have a well-defined beginning and end, and are generally well understood.

There are many designs that have a need for transactors that don't naturally fit into the processor model. Some interfaces are simply a continuous data stream. A low-level device may only see a stream of data. There is no start point or end point to it, even though the data stream almost always does have some structure that is visible to a higher-level function.

The example in Figure 4-6 shows a device that handles a stream of data. The data ingress transactor drives a continuous stream of data, while the egress transactor receives a data stream from the device under test. The processor is included in this figure only to show that a separate sequence of events may be needed to initialize or perform data transfers within the device under test.

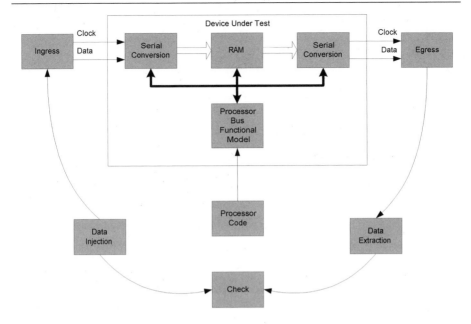

Figure 4-6: A Simple Data Transfer Example

Notice that in this example, multiple transactors must be operating in parallel. The data injection must be able to supply data to the ingress while, at the same time, the data extraction must be able to capture and return data.

In order to validate the extracted data stream, the checker module must know the stream that was sent in, and must then do a comparison of the two streams. This only works if the output stream looks identical to the input, and also requires the checker to be able to synchronize the streams in order to be able to compare the bits.

While conceptually simple, this type of mechanism has many serious drawbacks. Most devices do more than simply pass data through unaltered. They tend to modify data in some ways. Given that the data is a single stream of bits, the check routine would have a hard time verifying all but the simplest modifications.

For any realistic data set, the check function would in fact be a very complex piece of code that must essentially model the device under test. While that may be an acceptable solution, it does not provide the structure or modularity that can encourage re-use of components between environments.

Structure in Data Streams

In almost any data stream, there is some underlying structure. This structure may be as simple as a synchronization pulse that indicates word boundaries. Most communication devices use one of many protocols that include a great deal more structure. The type of data included in the structure has a great deal to do with the application. For example, many communications protocols specify headers such as source and destination addresses, as well as a payload, and an error-checking trailer such as CRC or checksum.

If the verification environment has knowledge of the data structures, it can take advantage of that information. The test environment has the ability to take advantage of four basic qualities that provide a structured, re-usable environment:

- Modularity

- Application interfaces

- Abstraction

- Encapsulation

This structure can be defined at several levels. Figure 4-7 shows an example environment where the data is structured as a packet with several defined fields: source and destination headers, some other potential headers, a payload, and a checksum.

In this example, there are two components to the test environment, the data generation and the results checking. They both benefit from a structured verification environment.

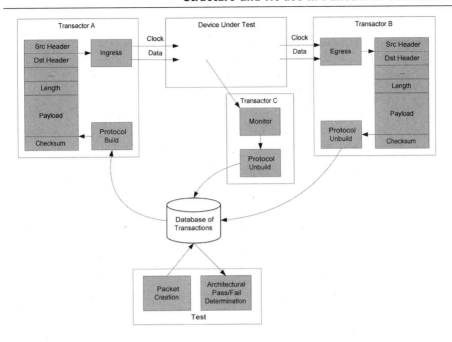

Figure 4-7: Structure in Data Streams

Data Generation

Data Generation Abstraction

The data generation part of a test environment is responsible for creating stimulus. With the structure provided by the transactors, the lower-level tasks of serializing data, packing headers into the data stream, driving data on the bus, and checking for protocol violations are all handled at the transaction level. The test can be written to concentrate only on the actual data desired. The test level is usually concerned with creating data fields necessary for conditions of interest in the device, and determining the type, number, and size of packets to be driven into the device under test. A good transactor should allow the test to easily control all the critical fields within the packet. Once that has been defined, other components of the test environment are responsible for converting the abstracted data to the particular protocol desired.

The test is abstracted from the protocol details, and need be aware of the fields only within the protocol that must be filled in, and how to create specific conditions such as protocol errors or timing delays in the packet transmission. If the protocol changed to a different, but similar protocol, the test may not need to be modified at all.

This also enables verification engineers of different levels to be able to work together effectively. As engineers start on a project, it is not necessary to understand all the details of the verification environment in order to effectively test parts of the architecture.

Interface from Test to Transactor

A well-defined interface between the data generation and the transactors allows any transactors of a similar ilk to use the identical API or standard interface methods. This makes the integration of a database simpler as well, since the database may likely use the same interface. Because the test is dealing with traffic issues and communicating through a standard API, it is possible for the test to be re-used between projects that have a similar architecture or in the lab. Just as important is that the transactors may be re-used or mixed with other transactors in other projects.

Transactor Modularity

All details of a particular protocol are contained within a transactor. While the details of the API may differ somewhat, based on the types of information that must be passed between the test and the transactor, all other protocol-specific functions are entirely encapsulated within the transactor. Neither the test nor the user need to be aware of any of the details of the protocol. This allows transactors to be re-used between projects, and it also eases the test writing task, since the test is able to concentrate on the device under test (DUT) rather than on the details of a particular protocol.

Data Checking and the Database of Transactions

The structure of the environment in Figure 4-7 permits both the test and the results checking to operate at a higher abstraction level. This permits complex checking functions to be done more easily.

Data Integrity

Generally, the first responsibility of results checking is determining if the data that was injected into the system made it through. This may be a straightforward process, since all injected transactions are visible in the database, and may be compared with any extracted transactions. Should any deviations occur, perhaps a data error, or missing data, the database can be used to determine what transactions were injected into the system, to aid in the debugging process.

Data Transformations

Many devices not only pass data through, but also modify the data in various ways. This makes the task of results checking somewhat more complex. As a simple example, a packet may have headers added or modified within the device under test. In this case, the extracted data should contain a similar packet, perhaps with additional headers.

Verifying the operation of the device under test is still quite manageable, since the database will contain all the components of the injected and extracted packets in a structural form. Each header can be compared individually. Determining that added headers are correct requires that the test writer understand what the device under test is expected to do. Finally, checksums are likely to be altered as well, and must also be verified. Frequently a separate function is used to verify checksums automatically.

Monitoring at Multiple Points

Figure 4-7 also shows a monitoring transactor. This is a transactor that is able to extract data, but does not interact with the device under test. It is responsible for listening on a particular protocol without interfering with the data flow on the bus. This may be a complete transactor that has its drivers disabled. A normal transactor would not just monitor, but would also actively drive responses on the bus. When a transactor is to be used as a monitor, then it is expected that there is already an active master and slave on the bus. In this case, the transactor can be used as a monitor, but must not drive any signals on the bus. Instead, it will receive all signals, and report any inconsistencies in the protocol.

This monitor provides the same encapsulation, modularity, and interface of any other transactor. The results can also be stored in a database of transactions. The monitor is used to provide additional information to the results checking routine, and may potentially aid in the debugging effort.

For debug purposes, a monitor can indicate when a particular transaction was seen at the intermediate point, and determine if an error had already been introduced. It can also indicate if problems are expected. Examples might include high traffic levels, or filled queues.

Performance Analysis

Many complex systems today are focused on high-performance applications. As a result, it is important for functional verification to analyze performance metrics as well as data integrity issues.

There are many potential aspects to performance analysis beyond the straightforward throughput metrics. Hardware architecture is often designed for complex decision-making in an attempt to improve overall performance. Examples of this include scheduling algorithms focused on choosing the optimal input to process next, or network algorithms that make queuing and packet-dropping choices to meet quality of service goals during times of high traffic.

For all of these examples, a significant amount of hardware complexity has been introduced in an attempt to optimize system performance. Functional verification must be able to inject and measure traffic to determine that the correct operations are taking place. In many cases, the data may pass through a system just fine even though the performance-based hardware has failed. In an example from the previous paragraph, if a network algorithm chooses a non-optimal incoming packet first, both packets will still go through the system, but the performance will not be what was expected. Using intermediate monitors and a database is one way to track transactions as they pass through a system of complex blocks.

Figure 4-8 shows an example system where transactions are transformed as they pass through a system. Individual packets are aggregated into more complex packets. These packets may be T1 lines coming in, and Sonet frames coming out, for example. They may also have

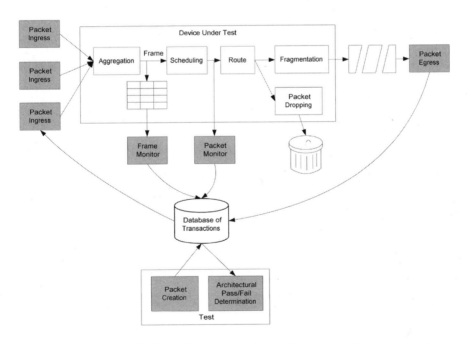

Figure 4-8: Data Transformations Through a System

been encapsulated into another format. As they pass through a network, other modifications may be made, such as fragmentation, packet dropping, or packet reordering to achieve quality of service goals.

Tracking transactions through this type of system to determine both functionality and performance may well require examining how each block individually reacts with the total system. Each block may operate differently in a system context than it would in a stand-alone environment. Since interactions between blocks are often critical to performance decisions, and the feedback from the other blocks is missing, it is often difficult to determine how well blocks will operate together by looking at them only in a stand-alone environment. However, by injecting data into a system environment and examining how all blocks work together, it is often possible to understand the performance issues of the complete system.

Critical to this is a database that stores how individual transactions migrate through the system, and allows this information to be examined for the thousands of transactions that are often necessary to create the traffic problems.

Transactor Layering

Some complex protocols are made up of multiple layers. This is quite common in communication protocols, but also exists elsewhere.
In these cases, it is often useful to have multiple layers of transactors to build each layer of the protocol. This maintains modularity and also allows intermediate results to be used and examined.

Figure 4-9 illustrates how a packet frame protocol such as Sonet can be built up from an intermediate packet type, which may itself be composed of several sources of packets, each using a virtual port.

Transactor layering may also be useful for examining intermediate points in a design. It is possible to deconstruct a complex packet into

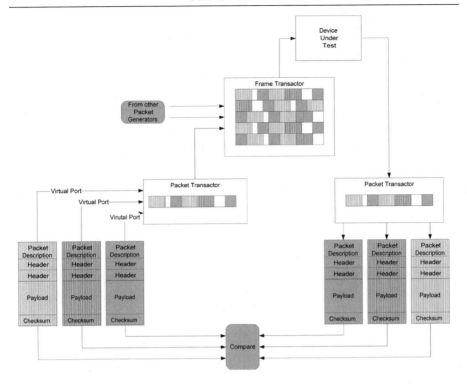

Figure 4-9: Layered Transactors

the fundamental pieces, which may then be compared with the original input. This can happen at any point in a design. An example of this was shown in Figure 4-8, where the intermediate frame monitor can break the frame back down into individual packets to be compared with the original incoming packets.

Re-use of Verification

Building a verification environment and the associated tests is a highly time-consuming process. Most project reports indicate that between 40% and 70% of the entire effort of a project is spent on verification, with 70% being much closer to the normal level for successful projects. This high level of effort indicates that the potential gains to be made with successful re-use are significant.

As stated earlier, a structured verification environment may permit many components to be re-used. Ideally, one would like to re-use as many components as possible, including the tests, the transactors, and any other environment functions that may be needed in a project.

Re-use may occur both within a project and between projects. There are some significant differences in how components are re-used within a single project, so that will be discussed on its own.

Re-use Within a Project

For any new project, verification usually starts at the block level. As the project progresses, and blocks are integrated into a system, a very different environment emerges that is larger, and more complex. There is a potential to re-use verification components between these phases.

Block-level Verification

Most projects do not start with a complete set of hardware designs available for functional verification. Usually a design comes together as smaller blocks. Then, the blocks are integrated into larger blocks, which may eventually be integrated into a system. There are several major reasons for performing functional verification at a block level.

Start Sooner

Because blocks of a system tend to be available before all of the design is done, block-level verification can be started as soon as a few blocks are available. Given the long functional verification cycle of a project, the sooner verification is started the better.

Access to Internal Functions

When a verification environment is able to connect directly to the inputs and outputs of a block, it is often easier to test some of the corner cases of the block, or to test a complete range of inputs. With a direct connection to the inputs of the block, a test should be able to specify exactly what is to be driven on that block. In contrast, when a particular block of logic is in a larger system design, driving the inputs of that block often involves programming other blocks so that they will pass the appropriate information onto the internal block. This indirection complicates the test writing and debugging tasks.

A disadvantage that must be considered is that internal interfaces are likely to change more frequently than system interfaces. This may happen when an interface was not fully defined, or when some unexpected additional functionality is required. When this happens, it will affect the transactors that are connected directly to block interfaces.

Faster Simulation
Individual blocks will be significantly smaller, often ranging from ten thousand to one-hundred thousand gates. Without the overhead of the complete system design, the verification tools are able to run much faster, which allows a more comprehensive set of tests to be run in a reasonable time frame.

Block-to-System Migration
In order to re-use verification components between the block and system level, the components need to be structured so that they will be able to operate in different environments. As always, modularity, encapsulation, good interfaces, and abstraction will be important. These features will allow a building block approach, where the same blocks in different configurations may be used as the verification environment is migrated from block to system.

The building blocks may consist of transactors, intermediate functions, such as the build or unbuild functions shown in some earlier diagrams, and some of the tests themselves.

Not all components will migrate from a block- to a system-level environment. For example, some block-level tests will concentrate on particular corner cases of a particular block. This type of test will generally remain in the block-level environment.

In order to re-use verification components between the block and system level, it is important to plan the structure of both environments so that components can be shared. Since different test levels have different requirements, modularity of shared components allows each level to re-use the components in different environments. Since this is good programming practice anyway, planning for a re-used, shared environment tends to result in a higher quality environment that saves time in the project even for blocks that aren't re-used.

The typical functional verification tests that are done on blocks overlap somewhat with the tests that are needed in larger integration systems. Test-level abstraction is critical to test portability. Tests that are written to concentrate on delivering specific traffic patterns and that are properly abstracted are well suited to working in both block-level and system-level environments. Project test plans are important to understand which tests should migrate from the block to the system environment before the test writing starts.

Figure 4-10 shows components of an example shown in Figure 4-8. The same verification components that are used in the system-level verification environment are suitable for the block-level environment as well.

Other blocks of the design may be verified in the same way, as shown in Figures 4-11 and 4-12. One point to note is that a transactor that is used in a block-level environment may become a monitor at a system level. Once blocks have been connected to each other, the blocks will provide the active interfaces, but the transactor can still be used to collect information as a monitor that not only checks for protocol

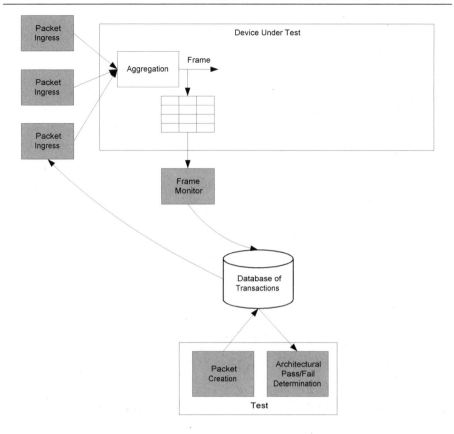

Figure 4-10: Block and System Verification Share Components

violations, but also captures and records all passing transactions into the database of transactions.

The ability to successfully re-use the verification components in different phases of a project requires that the components be designed to be modular. The functions of each block must be clearly identified, with well-defined interfaces.

It is important to ensure that block-level components will fit into a system-level verification environment. There are many pieces that will need to be assembled as the environment is migrated into multi-block and system level. Without planning, the pieces are not likely to fit together well. The structure of both the block- and system-level

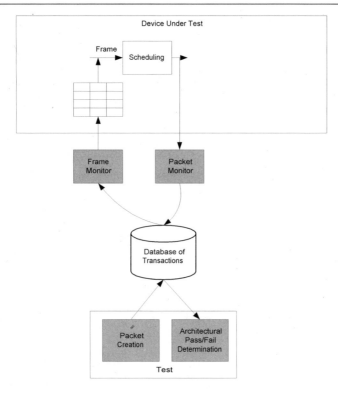

Figure 4-11: Block and System Verification Share Components

verification environments must be structured so that there is a high degree of overlap of transactors and tests.

Where sometimes block- and system-level verification environments are completely separate entities, it is possible to have a single environment that uses many of the same components and tests throughout the project.

Careful planning is needed to ensure that the block- and system-level environments will be based on the same verification architecture. The development of the blocks should occur after the verification environment is specified so that the components can be shared between environments. For further discussion on verification specifications, see Chapter 9.

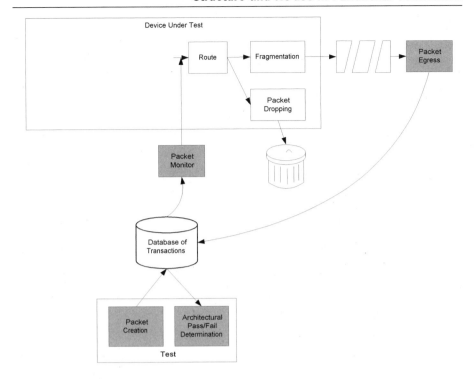

Figure 4-12: Block and System Verification Share Components

Re-use Between Projects

Re-use between projects is generally different from intra-project re-use, in that too frequently it is first thought about when the second project starts. It is rare for project-level re-use to be planned or anticipated at the start of a first project, since so much focus tends to be placed on the immediate project. More often, it is at the start of a subsequent project that project-level re-use is first considered. This is one reason that it is not unusual to have each new project develop a completely new verification environment.

One of the greatest challenges of re-using any type of software code is determining which pieces of code do what, and how to extract the useful components. Because this is often a long complex task, existing environments are often abandoned, with the justification, often correct, that it will be easier to write a new verification environment than to find the useful pieces of an old one.

One of the primary advantages of a structured, modular verification environment is that the environment is composed of individual components whose function can be identified. This means that it should have a well-defined interface to other components, perform a single function and encapsulate the knowledge associated with that function, and be as self-contained as possible so that the function can be extracted for re-use in another project.

With a well-structured environment and reasonable levels of documentation, project-level re-use can be quite effective. This is true for RTL design as well as functional verification. Large projects that have some similarities in architecture can achieve significant levels of re-use of verification code. Since verification is such an expensive part of the overall project cost, re-use tends to be important for project productivity.

Goals for Re-use Between Projects

Starting from an existing structured verification environment, new projects can achieve better than 50% re-use levels of all the verification code, which includes tests, transactors and intermediate functions.

One of the reasons that such significant re-use targets can be achieved is that projects tend to be related to each other. They may be different implementations of the same architectural family, or they may include earlier blocks of design code along with some new components that are integrated into the next generation.

It is common in these situations to re-use a number of tests as well as the transactors and functions in the verification environment.

While it is rarely pleasant for engineers to contemplate this, documentation also plays a significant role in the ability to re-use code.

A module that includes information on the goals, intent, use model, and interfaces is much easier to understand and extract. With this type of information provided, much less reverse engineering is needed to understand the function of that particular module, and the likelihood of re-use increases.

Summary

Structure in verification environments is critical to dealing efficiently with complex projects. A well-structured verification environment has many advantages, including the ability to manage the interactions of the verification components and to ease the task of test writing, which are often the single most time-consuming aspect of a project.

Critical to a structured environment is a well-planned, well-documented set of modules that provide encapsulation of protocols—a consistent interface—so that higher layers of the protocol can be written in a more abstract fashion. This structured environment is not simple to create. It takes significant time and effort to design, write, and debug the layers. However, it is invaluable for managing the greater task of creating and running tests.

Re-use, both within and between projects can provide significant time and productivity improvements. Given the large cost of providing functional verification to a project, effective re-use of verification components and tests is becoming critical to controlling overall project costs and schedules.

Random Testing

Key Objectives

- Achieving randomization

- Structuring random tests

- Constrained randomization

- Limits of random testing

Tests without randomization, often referred to as directed tests, are used to create a specific set of conditions that will be verified. A directed test is generally effective at creating the specific condition it was written for, but it does very little else. There are two basic limitations with directed tests. First, there are usually a large number of boundary conditions in any reasonably sized project. Often, there are more conditions than one could ever afford to write tests against. Second, for anything except the most trivial designs, not all boundary conditions are easily known. Without knowing all critical points of a design, it is not possible to write directed tests specifically to examine all critical parts of a design, even if there were sufficient resources available. Randomization is used in an attempt to overcome both of these issues.

Randomization can be used to explore more of the state space of a design, so that the test is able to exercise a larger percentage of the boundary conditions in the design, including many of those that may not have been identified.

Achieving Randomization

Randomization can be introduced at many levels. The most basic randomization occurs at the lowest level of a test where it interacts with the hardware. A low-level test may randomize the timing between simulation events, such as the delay between the assertion of a request signal and the resulting grant. In some systems it may be possible to randomize the data path values that are passed through the design. A memory test is an example where the data values often do not affect the control flow of the test, and may therefore be randomized with little effect on the test itself.

In a structured or transaction-based verification environment, randomization can occur within each individual transaction, as well as between transactions. Randomization within transactions may be structurally hidden from a higher-level test. This allows for abstraction of randomization similar to test abstraction, as was discussed in Chapter 4.

Transaction Randomization

Within a single transaction, there can be several potential areas for randomization. Figure 5-1 illustrates a simple processor-like operation, and the points where one could introduce randomization in a single transactor.

In this example, there are two types of random modifications being shown: those that affect the timing of the transaction, and those that affect the data being passed through the transactor. Some transactors may also be able to randomize the status and error conditions to cause more complex interactions.

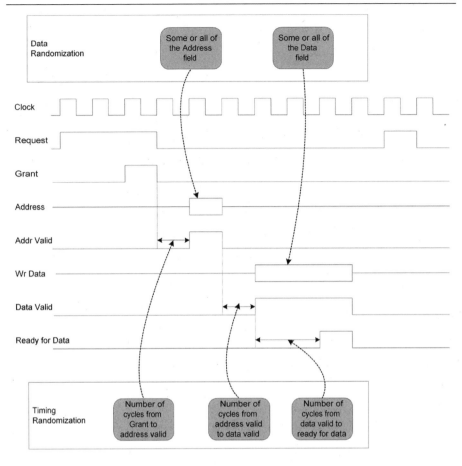

Figure 5-1: Randomization Within a Transaction

Data Path Randomization

Randomization of data in a system may be straightforward when the data does not directly affect the system under test, as in the previous memory test example, or for systems where data is simply passed through without affecting the control portions of the device.

When the data path does not affect the control of the system, the randomization is clear-cut. The test is affected only to the extent that it must know what the random values were so that they can be checked at some later time.

In more complex situations, the randomization will affect the control flow of the system. A simple example of this is in address randomization in a processor-based system. For addresses to be randomized, the test must generally have a way of understanding the effects of the randomization. First, it must be able to choose a legal random value that will address some part of the system, and second it must have a way to determine whether the results of the random operation are correct.

In the case where a test must determine the effects of randomization, the test must be significantly more complex, since it must be able to predict the outcome of the operation. In many cases, the test uses a predictive model of the device under test, or some other external aid to determine if the device under test acted appropriately.

Timing Randomization

There are many places where an architecture permits a variable number of cycle delays. As a result, timing randomization usually refers to multi-cycle delay randomization. Some examples of timing randomization that may occur within a single transaction are shown in Figure 5-1.

Randomization may also occur between transactions. For example, a processor test may introduce random timing delays between processor operations. This can occur in two ways. One is that an arbiter is programmed to delay between the assertion of request and grant, which effectively stalls the bus. Alternatively, the processor test could specify that a random number of no-operations (NOPs), or idle cycles, are to be inserted between read and write commands.

With timing randomization only, it is possible to create more diverse stimuli for a system under test. Figure 5-2 shows a very simple case of two processors issuing operations on a shared bus. Without any randomization, the bus operations are orderly and predictable. In this example, tests A and B could be two instances of the same test.

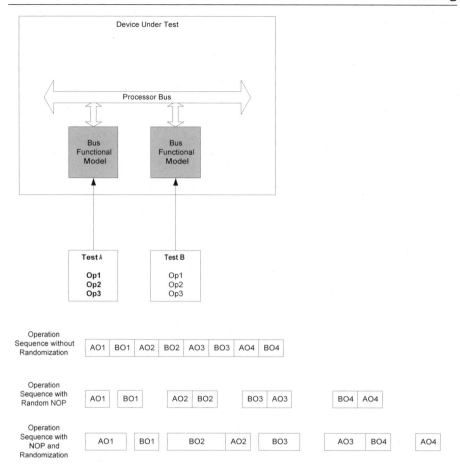

Figure 5-2: Randomization with Timing and NOP Insertion

Introducing timing randomization in the transactor can impact the order of operations and the interactions that must occur both on the shared bus as well as in the system under test. With the addition of random NOP insertions, a wide range of operation sequences can be achieved.

What should be noted is that a straightforward test sequence can be changed into a much more diverse set of bus operations with the introduction of a few simple randomizations within a transactor.

This can be done with minimal impact on a test, and with a fairly low amount of additional programming effort. It is the ability to create a range of new operation sequences with little additional overhead that provides much of the power of randomization.

Sequence Randomization

When there are multiple tests running, it may be possible to introduce randomization in the sequence of operations that are driven into the system under test. The purpose of sequence randomization is to interleave the operations of various tests to inject a more complex and realistic set of operations into the system under test. In Figure 5-3, a bus functional model is shown that is used to run multiple tests. In this example, there are multiple instantiations of only two tests, A and B. This example would be equally valid if this were done with six different tests. An interleave function can be used to randomly select which test to choose an operation from. This allows the sequence of bus operations to be varied, much as they might be in a multi-tasking operating system. For the tests in this example to work, each test must have either an independent set of resources, such as a unique range of memory to operate on, or a mechanism to share resources between tests without interfering with other tests.

Again, it is possible to write tests that are each straightforward, but with a combination of randomization methods that can create a wide range of sequences to stimulate the system under test.

Control Randomization

Another way to add randomization to a test is to randomize the control values in the device under test or in the testbench. By changing some basic values, the sequences of events can change significantly.

Within a device under test, there are generally timers and other control registers that determine when certain operations take place.

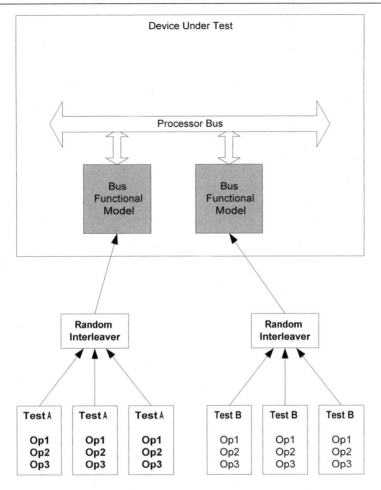

Figure 5-3: Sequence Randomization

One example of this is a refresh control register for a dynamic memory controller. Randomizing the refresh rate of the memory may change not only the sequence of operations on a memory bus, but also the delays and interactions between various operations as the refresh sequences interfere with other transactions. Other examples may include bus arbitration modes or priority settings.

Generally there are also controls that can be randomized in a testbench. Wherever there are multiple clocks, a slight change in clock frequencies will cause the interactions between the clock domains to change. This will not only help test the areas that connect different clock domains, but will also cause differences in the sequence of operations within each test range.

Control randomization should generally be viewed as an additional way to find different sequences of operations, rather than as a primary source of randomization. While it will cause some differences in sequence interaction, it will not affect the sequences themselves, and this is the primary source of randomization.

Random Test Structures

Randomization is sometimes characterized as being either directed random or random directed. The difference between these two methods lies in the way the underlying test has been written.

Random Directed Tests

Random directed tests often come into being as pure directed tests to which randomization has been added. They are also useful when specific sequences of operations are necessary, and randomization is only added in specific areas. The previous examples are random directed tests.

Directed Random Tests

A test that is structured as a directed test which has randomization added is often referred to as a random directed test. Another possibility is to create a random test and then provide some level of control on the randomization so that the test performs the desired operations. This is a directed random test. Unfortunately, there are many different names for various testing methods, and those names are not always constant.

Directed random tests can be thought of as creating a distribution of orthogonal fields for each new transaction. The test will create a new transaction, and the randomization is responsible for filling in all possible values for the transaction.

Bag Randomization

Bag randomization is sometimes used to create directed random tests. When specific sequences aren't required, a bag randomization can be used to create random streams of operations. In this case there are many potential operations that can be chosen. All of these are available and they will be randomly picked out of the bag one at a time. This is referred to as a bag algorithm, and is shown in Figure 5-4. Each operation will be executed before any operation is chosen twice. Alternatively, operations may be weighted to increase or decrease the probability of each operation being chosen.

Notice that the bag algorithm will result in a more random sequence of operations than interleaving a series of tests. In some ways, more randomization may be useful, since it might create a different command sequence than the interleaved instructions. On the other hand, randomized tests as shown in Figure 5-4 are able to perform test sequences that operate on the device under test. The semaphore test that was shown in Figure 4-4 is an example where a sequence of operations is necessary in order to test device functionality. A simple bag of operations is not sufficient to create such a sequence.

Mixing Directed Random and Random Directed Testing

Since both sequences and bags may be useful, it is common to mix both directed random and random directed tests together. There are many ways to accomplish this. One way may be to use a multi-level bag algorithm, as shown in Figure 5-5.

This is just one way to structure this type of approach. The key step to any multi-level randomization is to find a way to divide the decision

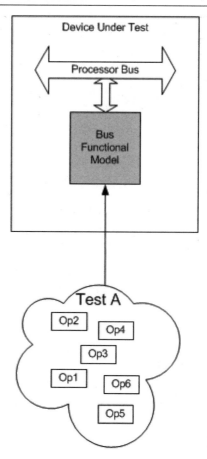

Figure 5-4: Bag Randomization

process into multiple pieces. In this example, the first step is to decide whether to run a test or a random operation.

Constrained Randomization

If randomization is so wonderful, why not just randomize everything and wait for the errors to fall out all on their own? This is sometimes called the shotgun approach to randomization.

Simply randomizing everything and running a long simulation is not likely to be the most effective way of finding boundary conditions.

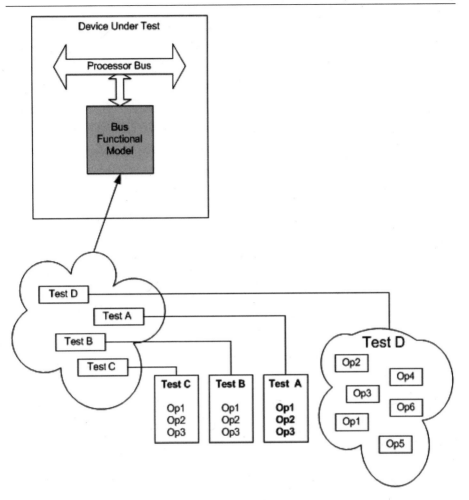

Figure 5-5: Multi-level Bag Algorithm

A purely random test has not been provided with any information to concentrate on the boundary conditions. As a result, it is likely to test throughout the design space. While that sounds reasonable, it is generally impractical. There are a few reasons for this. No simulation is able to cover most of the state space of a design. A reasonably complex design will consist of perhaps one million gates, with perhaps 50,000 registers. With a total state space of $2^{50,000}$, and a simulation speed of perhaps 1000 cycles per second, even assuming that a unique state can be reached on every simulation cycle, this is far beyond any practical timescale.

Simulations can still be effective because boundary conditions tend not to be in random places. An error is far more likely to occur at endpoints, during exceptions, or at the limits of any given range.

As a result, a purely random simulation is not likely to be effective. Some mechanism to focus the simulation near potential boundary conditions is required. One way to do this is by applying constraints to the randomization. This can be used to aim the randomization towards areas that are more likely to contain interesting boundary conditions.

Commonly referred to as constrained randomization, the purpose is to provide guidelines that will help steer the random tests to more interesting cases, and remain within the bounds of legal cases.

A simple example of this would be testing a counter. Much of the logic within the counter, and presumably some of the surrounding logic, is likely to be focused on the endpoints of that counter. This may be due to full or empty conditions, or overflows and underflows that the counter detects. A constrained random test may specify that the counter should be loaded with values near an endpoint in an attempt to emphasize the more interesting cases.

Constraints may also be used to specify the legal limits of certain values. This could be used to ensure that only legal op-codes are created for a processor test. Other constraints can be used to keep memory access within legal limits.

How constraints are specified varies greatly, based on the verification tools used for the project. Generally, a set of constraints will be used each time a new transaction is created. As the number of constraints increases, each new transaction will be more focused on a particular area.

To show how constraints work, consider an example where a processor must provide an address that is within a legal range. Figure 5-6 shows

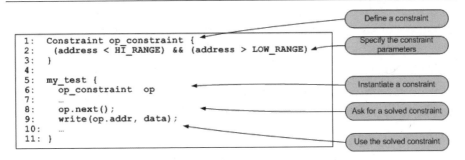

```
1:   Constraint op_constraint {
2:      (address < HI_RANGE) && (address > LOW_RANGE)
3:   }
4:
5:   my_test {
6:      op_constraint  op
7:      ...
8:      op.next();
9:      write(op.addr, data);
10:     ...
11:  }
```

Define a constraint

Specify the constraint parameters

Instantiate a constraint

Ask for a solved constraint

Use the solved constraint

Figure 5-6: Random Constraints

an example of how constraints can work. Note that this is stylized. Each verification language will have its own constructs.

Constraints are often used in conjunction with other randomized tests. They may be used to specify ranges, as well as to specify relations between different values. For example, a particular type of operation may be permitted a specific address range. A constraint solver can be used to pick a random operation, and then solve other constraints based on the chosen operation.

Constraints can become quite complex, as more relationships are added to the constraint equations. In some cases, the constraint specification may be one of the more complex aspects of a verification test.

Why Is Randomization Effective?

Random sequences are used in an attempt to find boundary conditions in a design that are unknown, and therefore unlikely to be discovered using directed tests. There are many sources of unknown boundary conditions. Many are caused by the interaction of multiple control blocks. It is often possible to determine all legal states and conditions within a single control block. Once two or more such blocks interact, it may be much more difficult to find all possible states of all the combined blocks.

When a directed test is run, it will create a single sequence of events that generally occur at very specific times. If the directed test is run multiple times, the same sequence of events will occur each time with no modifications. Randomization can be used to alter many things. It may alter the timing of the events, which would alter some interactions among state machines. Altering the sequence of events will cause a new and different set of interactions in the system, as will modifying the events themselves.

Randomization allows a test writer to create a much more diverse set of conditions and interactions that are likely to exercise more of the state space of the device under test.

It is difficult to measure the effectiveness of random versus directed tests. Some rough estimates can be obtained by looking at empirical data from a number of projects that used both randomized and directed tests. The number of design flaws found using each method indicates that randomization can be an effective method to detect flaws in a device under test. In some projects, random testing was able to find two to three orders of magnitude more design flaws than directed testing, even with fewer people engaged in the writing of the random tests.

Distribution Models

When using any type of randomization, it may be important to know how the random numbers are created. There are many different types of pseudo-random number generators, and their quality varies drastically. Some well-known random number generators are far from random, and even have bits that simply toggle. In addition to the quality of the random numbers, the distribution may also be important.

Different distributions are useful to mimic the types of traffic that might be seen in a particular environment. An appropriate distribution model will allow the functional verification to more closely mimic the environment in which the design will be used. This increases the likelihood that the design will work as predicted by the verification effort.

Many different distribution algorithms have been used for modeling various systems. A common one is the Poisson distribution. It provides a probability distribution of events over an interval of time that is independent of occurrences before that time interval. For event arrival modeling, Markov chains are also sometimes used to model systems that wait for events such as phone switches. A wide variety of distribution models have been created to mimic different types of real-world events.

The choice of distribution may be important if the traffic load affects the functionality or performance of the system. For example, queuing systems tend to be quite sensitive to the distribution of events. Small changes in the input or output distributions may have significant impact on the behavior or performance of the system. When the distribution model is important, care must be taken, since picking the correct model can be difficult. The easy part is understanding the various distribution models. The hard part is understanding the expected traffic distributions that the system will face.

Random Distribution Tools

In addition to the types of distributions, many verification languages support various mechanisms to choose subsets of random numbers. These can be convenient for either limiting test parameters, or for ensuring that all possible combinations are hit.

Range tools will ensure that all random numbers are within a specific range. In some cases, the distribution of the range may also be specified, in others, the type of distribution is not specified and cannot be altered.

When all possible values need to be used, a bag tool may be provided. The concept is that all possible values within a range are placed within the bag, and randomly picked. Each value is picked only once until all possible values have been used.

Random Seeds

Specifying a random seed can be helpful when a particular test needs to be repeated. The most common example of this is when a bug is found in a random simulation. Once the bug is fixed, the simulation needs to be rerun to verify the fix. To check if the bug has been fixed, the same random test must be run in exactly the same way so that the previous scenario will reoccur at a known time. Note that the fix to a bug may alter the sequence of operations enough so that the specific scenario no longer occurs. Because of this, it is difficult to ensure that the bug has been fixed without some additional effort. By adding specific tests, assertions, or monitors, it is usually possible to verify that the test checks for a specific bug.

In order to ensure that a random test is run exactly the same way, all the random numbers must be the same as in the original run. Pseudo-random number generators create a sequence of numbers starting from a seed number. If the same seed is used, then two sequences of numbers will create the same random numbers. From a verification viewpoint, it means that a random test run multiple times with the same seed will produce the same results. A good verification environment will display the random seed for any particular run, and permit that seed to be used for any subsequent run. Without this seed, it is difficult to determine the effects of a particular modification on a specific random test. By running a random test with the same seed after a fix has been made, it is usually possible to ensure that any problem detected in a particular run has been repaired, as long as the fix has not changed the random sequence in some way.

Most verification languages provide a number of convenient features to ease randomization. This may be an important criterion when choosing a verification language.

Since regression tests are usually run regularly, they provide an opportunity for randomization. This requires that each run use a different random seed. The advantage of having each regression test run with a different seed is that it permits a wider range of random values and sequences. While this can cause regression tests to fail on any run, that is one of the purposes of running tests.

Structuring Effective Randomization Tests

How a random test is built depends a great deal on the verification methodology used for a particular project. There is no single method, or even a best method for creating a random test. Below are constructs that could be used to create random tests in a structured transaction-oriented environment.

Timing Randomization

As was seen in earlier examples, timing randomization can be inserted into a transactor at a very low level of the test structure. Since this often does not affect the control flow of the test, or the device under test, it is possible to add timing randomization to a test without modifying the test at all.

A simple random test may look exactly like a normal directed test. Figure 5-7 shows a simple processor memory test that is a random directed test. Randomization may be added by a separate function that modifies timing, and potentially inserts idle blocks.

In this example, a directed test is used to define the operations. An intermediate randomization module is modifying the timing specifications of the test, and potentially introducing additional idle cycles by inserting NOPs.

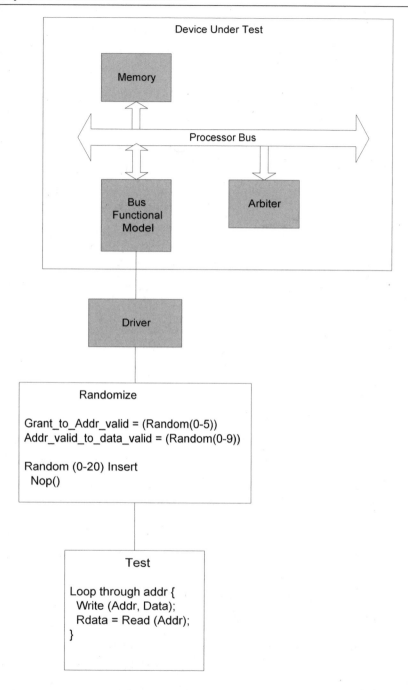

Figure 5-7: Timing Randomization on a Random Directed Test

By randomizing timing of a single test sequence, some new patterns will be generated. The processor bus is exercised somewhat better, since there are different operation lengths involved, and perhaps the arbiter as well. Overall however, by only randomizing timing with a single test, the transactions are usually driven in the same sequence.

In some situations, it may be important that the test not be modified. This could be because the test is actually driven from some other source, such as a separate machine, or it may be that the test is a piece of software that is used elsewhere, such as a driver or diagnostic meant for the real machine. In any of these cases, it is possible to take advantage of some of the power of randomization without affecting the test code.

Data Path Randomization

Processor tests are limited in what data can be randomized. For simple processor transactions, there are only three data fields, the address, the data, and the operation itself. When the functionality of the test must be maintained, at least some of these fields cannot be arbitrarily randomized. Instead, it may be important to determine which fields can be randomized in a way that does not affect the test functionality. In some cases, this can be simple. For example, a memory test is often not affected by random data, and possibly not even by randomizing a subset of the address field.

One method to provide more randomization while using directed tests is by selecting randomly from multiple concurrent tests. Figure 5-8 shows a set of processor tests that are all being run. Individually, each test may provide randomization within that specific test. The selection function that determines which operation to choose from allows the processor transactor to issue many different transaction sequences.

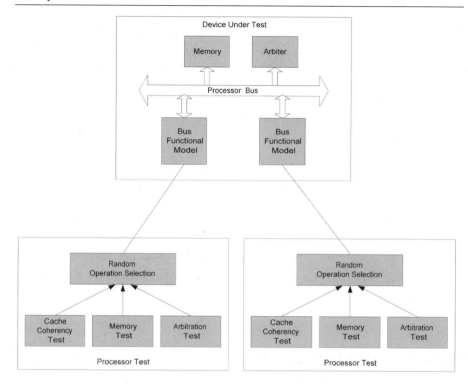

Figure 5-8: Sequence Randomization from Directed Tests

The result of this type of randomization is a wide range of timing, data path, and transaction sequence randomization, while each test maintains the functionality for which it was designed. The tests can still be written as self-contained self-checking tests. Some tests may be instantiated multiple times so that there are multiple concurrent instances of the same test. This is particularly effective when a mechanism is in place to allow each test to have a unique memory range so that multiple instances will not interfere with each other.

Data Traffic Randomization

As with processor operations, most data traffic operations can be randomized as well. If the data is never examined in the device under

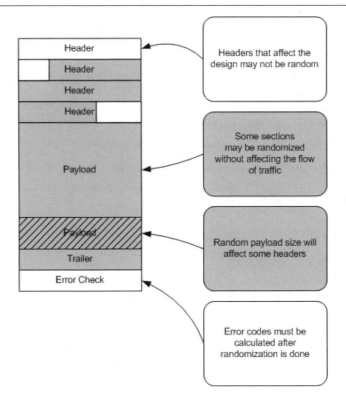

Figure 5-9: Randomization Within Data Traffic

test, then randomization is simple. In most cases, however, data traffic does have structure that is interpreted by the design. In that case, as with the processor operations, only portions of the payload can be modified so that the integrity of the test is not affected. Generally, portions of the data can be modified as shown in Figure 5-9. Some sections are not examined, so they can be fully randomized, while other headers are used within the device under test.

Randomization within a single stream of data is useful, since it will stress some of the design under test. Where multiple streams of data can be handled, the ability to interleave multiple data streams allows a wider range of stimulus, similar to randomizing sequences in processor tests.

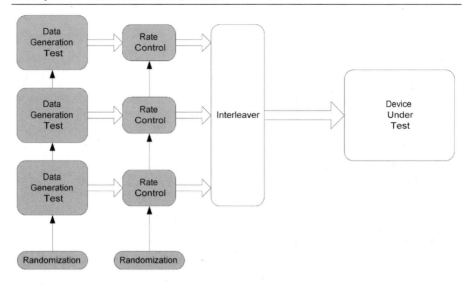

Figure 5-10: Data Traffic Randomization

As in the processor case, each data generation test need not be very different from simpler tests. Figure 5-10 illustrates the interleaving of various data from multiple tests.

Control Flow Randomization

In some cases, it is possible to write tests that randomize the control flow of the device under test. Such a test would provide randomized inputs that would affect the operation of a device under test. An example of this is a test of a processor that randomizes the instruction op-codes sent to the processor. By choosing a random op-code, the test is directly affecting the control flow of the device under test. For the op-code randomization to be effective, the choice of op-code should not be based upon the results of previous operations. Often, the most difficult aspect of control-flow randomization is in determining if the resulting behavior of the device under test is correct. In the op-code example, when a random op-code is chosen, not only does that affect the current instruction of the processor, but it may also impact the operation of future instructions.

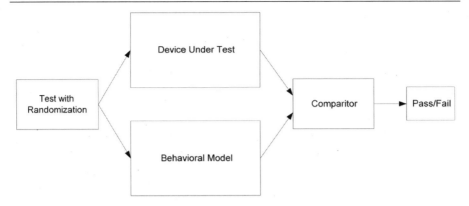

Figure 5-11: Reference Model Used to Predict Correct Behavior

This problem is sometimes broken into two parts: a test that provides the randomized stimulus, and a model that determines the correct behavior of the device under test. Figure 5-11 shows an example of this. The model may come from another source, such as a behavioral model that was created to aid the software development effort. It should be noted that developing and maintaining a reference model may be quite time-consuming, which is why this method is not always an optimal approach for some projects.

The Importance of Being Idle

While the generation of transactions is critical to a test, for randomization to be effective, one often-overlooked criterion is the absence of activity. Many devices are designed to handle a flow of data or operations up to a maximum rate, which is often less than the combined rate of all of the inputs to the device. As a result, if the input traffic is higher than the device can handle, queues will fill up and the slowest component in the device will regulate the throughput of the device. This may result in the verification environment providing results that are different from the behavior that a real system is likely to see, and may result in incorrect behavior or performance estimates.

If a test, or a set of tests, is driving sufficient volumes of traffic, the traffic will back up and be limited by the single slowest component in the pipeline. This results in a fairly regular flow of traffic being issued from the slowest component in a design. The output of that component is potentially much less random than the original traffic. As long as the rate of traffic being injected into the system is higher than the rate that a particular component can handle, then that component becomes the controlling part of the flow of traffic, rather than the randomized test. Much of the test timing randomization may be lost as this component drives the flow of traffic through the system.

The insertion of idle times, in random positions and random lengths, is an effective way to prevent this problem, and to control the transaction rates or data bandwidth into the device under test. In order to determine reasonable ranges of idle insertion, it is important to understand how tests are interacting with the device under test. In some cases, there should be sufficient traffic to fill queues. In other cases, interesting sequences may only occur when the queues have not been filled. Monitoring the device under test to find appropriate traffic levels is often necessary. This can also be done on-the-fly using feedback from the results in the database.

Idle times also allow for conditions to be hit that involve quiescent states. Without idles, the internal components will be busy at least part of the time, so the quiescent states are unlikely to be reached after the initial reset. This may cause a part of the state space not to be fully explored.

Determining how many idles to insert, and for how long, has a great deal to do with the architecture of the device under test and the type of testing being done. This generally requires some amount of examination and tuning to find a reasonable balance that allows for realistic traffic levels, and some quiescent periods.

Randomization and Test Re-use

Re-use in verification tests is quite common, and sometimes detrimental. When a new test is written for a project, it is rare that some code from previous tests isn't used. That code will have sequences that are re-used for many tests. This can reduce the test-writing effort but means that multiple tests will share common elements and patterns. It is not uncommon for multiple tests, which are concentrating on different areas of a design, all to have a very similar flow.

While test re-use is critical to the efficient construction of a verification suite, it may have a negative impact on the effectiveness of each test, since the tests are not independent of each other. The sequences of events are likely to be similar, and the result is that many of the same boundary conditions are likely to be exercised instead of finding new boundary conditions through new and different test sequences.

It is important to recognize the difference in coverage between independently written tests and those that share signification sections of code, since an independently written test is more likely to result in a different sequence of events in the device under test, which in turn is more likely to reveal a new and different issue in the design. This situation exists for any other shared test components such as drivers or algorithmic modules that are used repeatedly by the verification environment.

What this implies is that multiple tests, which are supposed to independently test different sections of logic, are likely to have some sections that are actually the same test. A group of tests that share sections of code will actually have a smaller distribution of event sequences than one might think.

Randomization in tests can mitigate this problem to some extent. Thorough randomization can create large deviations in test sequences, thus allowing good distribution of sequences as code fragments are re-used. If the test is carefully constructed, then re-use of the code is not detrimental, since it will result in a new sequence of events in each test.

Limits of Randomized Testing

There are some cases where randomization may provide overly optimistic or pessimistic results. This can happen when a random distribution provides superior performance to real-world traffic. One example where randomization is likely to be overly optimistic is in a design that contains a hashing algorithm. The boundary conditions of most hashing algorithms are stressed when the key fields are similar. This causes overflows of individual buckets, and prevents all buckets from being filled equally, causing inefficiencies in memory use. A random distribution, on the other hand, is likely to fill all buckets equally, resulting in an optimal hash table distribution.

Where the hashing key is generated from addresses, a random distribution of addresses is the best possible scenario for hashing. All buckets are likely to be filled equally, the chance of any bucket overflow is minimized, and any variable time delays are minimized. A pure random sequence may be the least likely to find any errors involving the boundary conditions between the hashing algorithm and the rest of the design.

If simulation metrics are being used to estimate the performance metrics of a system, then the choice of random distributions becomes important. A random distribution of addresses may be overly optimistic from a hashing perspective, but it would also be overly pessimistic from a caching perspective. Using a distribution that is similar to the expected distribution within a real system may be important for finding boundary conditions, and obtaining reasonably accurate performance metrics.

In most systems, traffic is usually far from random in nature. There are patterns in both the address space and in time. Processor addresses have a certain amount of locality, which is what makes caches effective. Data traffic headers may also not be random, since there are usually a number of connections that pass multiple packets of data.

Similarly, traffic is often not smoothly distributed in time. Processor and data traffic may both have bursts of activity at times, and be more quiescent in others. Since this represents the type of traffic that is likely to be seen in a real system, it is important for the functional verification to ensure that the system will perform correctly in those situations.

Since randomization may provide an overly optimistic set of stimuli, some high-stress conditions may not be triggered within the random simulation. These may be the conditions that are likely to be encountered in a production environment. Careful use of random values can overcome some of this, but generally some parameterization is required to focus the randomization.

Coverage Feedback

One frequently discussed technique in randomization is to measure the effectiveness of randomization and dynamically modify the randomization parameters to improve the test coverage, or determine test completion. As a random test is run on a device, the device is examined dynamically to see what conditions have or have not been covered. This information is then fed back to the test, which may then modify parameters in order to concentrate on areas that are not adequately covered.

Figure 5-12 shows an example of a directed random test that runs on a design. During the simulation, a coverage tool measures the functional coverage of the design, and provides feedback to the

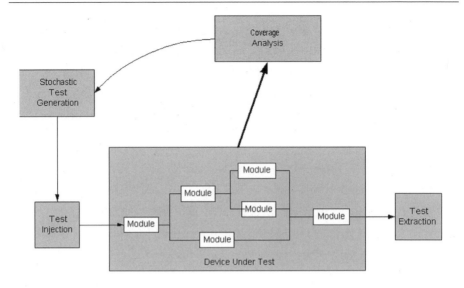

Figure 5-12: Coverage Feedback Loop

random stimulus. This method may be used to make some very specific measurements, or to determine if the test completion criteria have been met.

For test completion, a certain set of criteria may all need to have occurred. Once they have all been detected, then the test may complete. Similarly, a test may be expected to create a particular scenario. One example is a test that is expected to drive sufficient traffic to fill a queue to cause back-pressure through the system. The test may generate traffic until an external measurement indicates that the queue has been filled, relying on the external notification that the event has occurred.

These specific examples of feedback can work well. The coverage measurement is used for very specific types of information that the test can react to. The reason these types of examples work is that there are a few specific responses the test is programmed for. In these specific cases, the test is expecting feedback about a few predetermined criteria, and it is well understood which parameters need to be modified, based on the feedback that is received. It should be noted that this method works for non-random tests as well.

Limits of Coverage Feedback

There is considerable talk in the industry about the power of coverage feedback in handling more general cases. These would be cases where broader information of coverage analysis is used by the test to determine how to alter parameters to increase the functional coverage in any areas where a weakness is indicated.

There are several significant issues with more general cases. Code coverage metrics tend to provide low-level reports indicating which lines or paths of RTL code are not fully covered. It may be quite difficult to determine what any particular line of code does, and therefore the types of operations that need to be done in order to provide stimulus for that particular area. Functional coverage tools may provide a higher-level feedback mechanism, but translation from feedback to test parameter modifications is generally still needed.

The more general case is usually more difficult. This would be where a coverage analysis tool indicates a weakness in coverage, and the test uses that information to strengthen the coverage in that specific area. One issue with this is that the coverage metrics are often quite low-level, indicating the occurrence of some very specific events. It may be quite difficult to determine how to improve a specific set of coverage metrics by modifying the parameters of a random test.

This can sometimes be achieved at a very broad level by changing the ratio of tests or operations. On the other hand, for reasonably complex designs it is often not clear how a particular set of parameters will affect the coverage metrics in a system.

Effective random testing is quite complex. It depends on a set of well-constructed parameters to ensure that the tests are stressing the design well. Ensuring that the parameters are improved through the use of automated feedback is considerably more difficult than is sometimes claimed within the industry. It seems that no verification presentation is complete without a slide demonstrating the use of automated feedback with a random test environment. This is usually shown as a simple, nearly self-explanatory function. It is important to understand the scope and difficulties of this problem, since the claims about this technique may be beyond what is generally practical.

Running a Random Test

Before a random test is written, the goals of the test should be well defined, and the mechanisms that the test will use to cover those goals needs to be understood. One of the basic questions is how to determine the completion criteria of a test. How long should a test run, and how does one determine if the goals of a test have been met? Generally, some type of measurement is needed to ensure that specific events occurred as the test was run. While internal or external measurements may indicate that specific conditions did occur, random tests tend to have more than a few specific goals. Running the test for too short a time means that the test does not have a chance to hit some potential boundary conditions, while running it too long may be a waste of resources.

There is rarely a single answer to the question of test completion. One method to determine how long a test should run is to measure the number of problems it finds as it runs. Measuring the number of bugs that a test finds in a given simulation time is one way to measure effectiveness. If few bugs are found, this may indicate a problem in the approach of the test. On the other hand, if a test continues to find bugs, this is a pretty good indication that it should continue to run. A track record of test effectiveness may be one way to determine how long tests should be run, and which hardware areas and test areas may need to be improved upon.

Summary

While random testing is usually an essential component of any verification approach, it will be more effective with careful planning. Random testing can be quite expensive. A shotgun approach will require large amounts of simulation time, and may not be effective or productive. Similarly, limited randomization may not be able to explore the design under test sufficiently to be effective, and thus may not find all of the potential bugs that a better-planned designed random test could.

To maximize the likely effectiveness, planning is required to determine which areas of the design should be focused on, what attributes should be randomized, and the constraints and distributions of randomization. Understanding the expected results and the completion criteria for the test will also aid in improving the effectiveness and productivity of each test.

Co-simulation

Key Objectives

- Verifying hardware and software

- Methods used for co-simulation

- Choices based on project requirements

Most modern digital systems are designed with a mix of hardware and software components. Part of designing the architecture of a system is to specify how any particular system functionality is to be implemented: as a hardware or a software block. In many cases, the tradeoffs between these choices are not immediately clear, and any number of components may be switched from one to the other, based on a combination of performance, cost, schedule, or size considerations.

In any event, for the blocks to operate together as a complete system, both the hardware and the software components need to be present and functioning. As has been discussed in Chapters 3 and 4, high-level verification requires that these blocks be connected together to verify that they function properly. This is true regardless of the implementation of any particular block. Co-simulation is used to tie the hardware and software components of a system together. Many different types of software may be incorporated into a system, ranging from embedded micro-code in a processor development project to a complete operating system as might be found in a personal digital assistant. In the case of micro-code and other low-level software, it is a combination of hardware and software that makes a particular block function correctly. For operating systems and some driver code, the software runs on top of a hardware layer in order to perform a separate system function. In either case, the hardware and software are both separate functions that together, implement a part of a complete system architecture.

While the definition of the term may vary, co-simulation may be anything from a single snippet of code in an application, to a single driver, all the way to a complete operating system. The software will be interacting with a simulation of the hardware design. This may also vary from a completely simulated hardware system to an individual block that is being simulated while the rest of the system is implemented in hardware.

As system-on-a-chip (SoC) designs become more prevalent, the mesh between hardware and software components in a design becomes even tighter. Many of these designs are built almost entirely from purchased intellectual property, and have very tight schedule constraints. This means that the major tasks in the project are stitching the components together, integrating software, and verifying the functionality and performance of the entire hardware and software system.

A typical SoC architecture is shown in Figure 6-1. The device contains a microprocessor and the necessary support logic for the processor to run the applications internally, along with the functional blocks needed for the specific application the device is designed for.

While many people define SoC to mean that a microprocessor is a component of the system, there are quite a few designs that are similar in their functionality, but use an external processor. Figure 6-2 shows an example of this. While this type of system may not be considered to be a true SoC, the design and verification issues are very similar.

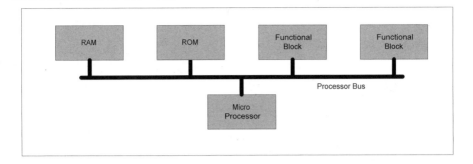

Figure 6-1: Classic System-on-a-Chip Architecture

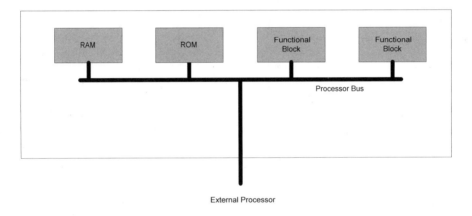

Figure 6-2: Chip Architecture with External Processor

In both of these systems, when some of the functional blocks are purchased, the project time spent in the design phase of a project may be significantly reduced, despite the fact that the complexity of the chip is still very high. While the block-level verification load should be low, since a purchased functional block should have been verified, the same may not be true at a system level. If the blocks are going to be used in a different way than that for which they have been previously tested, the probability of finding bugs in an existing, tested design is fairly high. This puts a great deal of pressure on the system-level verification of the design, since it may be the single largest time and cost component of the project. This must be accounted for in the functional verification plan.

Goals of Co-simulation

Co-simulation is a method of incorporating both hardware and software components into a single functional system that behaves in the same way as it would in a real implementation of the system. There are several common goals of co-simulation.

From an architectural viewpoint, most SoC projects look quite different than they do from a hardware implementation viewpoint. An SoC-based architecture is often meant to fulfill a specific purpose. This usually requires many specialized functional blocks that are able to manipulate or store data, and pass it on to other blocks. Examples of this might be cell phones, or a digital camera. Each of these must receive, analyze, convert, and perhaps store data.

From an architectural viewpoint, a system contains a series of blocks, each of which perform a specific function. By combining the functions, the system is able to meet the architectural requirements. At this level, the actual implementation of the block may be completely unimportant. Some blocks may be implemented in hardware, while others consist of an algorithm that is executed in software. The details of how they are connected together is dealt with at the implementation level.

Figure 6-3 shows what the flow of an architectural block diagram might look like. Notice that this type of diagram could be for the same device as the hardware view shown in Figure 6-1. Where Figure 6-1 shows a hardware diagram of a device, Figure 6-3 shows an architectural level. It is quite common at this level for the block implementation to be completely unknown. The block may later be implemented in hardware or software.

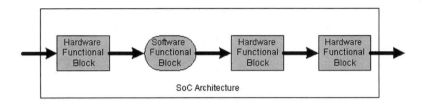

Figure 6-3: An Architectural Block Diagram of an SoC Device

Determine Architectural Validity

When a system architecture is defined by both hardware and software components, often the earliest time an architecture can be modeled is when there is an implementation of both the hardware and software. In some circumstances, determining how well an architecture fits its requirements can only be done with both sets of components. Co-simulation allows the full system to be analyzed to ensure that it meets the architectural intent of the system, and that all pieces perform as required.

Performance metrics are often a critical component of the architectural definition. Metrics may include system throughput, software latency, or chip size for example. Performance and cost metrics often require hardware and software components that are well matched to meet the requirements of the architecture. This is where some hardware/software tradeoffs are made. While software components are often less expensive to implement than hardware, they may not be fast enough. Similarly, hardware components can improve performance metrics, but they increase the power, size, and cost of the device. Determining that the correct functional blocks are implemented in hardware or software, and that there is sufficient processing power for the software, is part of what co-simulation can validate in a tradeoff analysis.

Verification of Implementation

As was discussed in Chapter 3, functional verification must often explore the boundaries of interconnected blocks. This is a potential source of design and architectural issues. In an SoC-based design, where some blocks may be implemented in software, co-simulation is necessary to explore the boundaries and interactions between blocks.

Verification of the implementation in a co-simulation model is similar to what happens in a pure-hardware verification model, with the exception that some of the blocks are designed to be implemented in software, and are written as such, while others are designed to be implemented in hardware, and are written in a hardware simulation language.

As in any other hardware implementation, boundaries and state dependencies between functional blocks are a potential source of design and architectural issues. By connecting them together and running functional verification tests, it is possible to identify and correct these issues at the simulation phase of a project. For both the hardware and software portions of the architecture, early identification of issues simplifies project progress greatly.

Co-simulation Provides a Realistic Verification Environment

Another advantage to co-simulation is often overlooked, but potentially just as important. For any reasonably complex system, it is not possible to verify the operation of a design for all sets of states and inputs. Therefore, for functional verification to be effective, it must attempt to create a subset of conditions that are close to those that the system will be exposed to in the real world. Where hardware must interact with a software layer, the closer the software is to the actual code that the device will be connected to, the more accurate and realistic the functional behavior will be. This means that the device is more likely to respond as expected when the system is built. As an example of this, consider a software reset sequence on a system. There may be some unexpected interactions between various blocks so that only some sequences work correctly. In this case, it is unlikely that all possible reset sequences can be tested. However, if the actual reset software is used, then the one reset sequence that really matters will be tested. In this obviously simple example, it could be argued that a different reset sequence can be found and used. For more complex logic, that may not be so simple to do.

Co-simulation Is Able to Share Software and Verification Development

When it is possible to use software to test a function, an equivalent verification test does not need to be developed. In the reset sequence example above, the actual software code can be used instead of developing a functional verification test. In many projects, where the software is not ready early enough, the inverse may also work. The reset verification test can be used as part of the software, rather than developing another block of software. Again, this is a simplified example, and not much effort will have been saved. However, for other components of a system, the savings of developing only one block of code may be significant. Even when the actual code is not shared because of the differences in the software and verification environments, it is still often possible to share the algorithms that were used.

Project Issues

In many project environments, the hardware and software teams are separate organizations, even when the final project requires that these two components be tightly coupled. Traditionally, these teams are coordinated only two times: at the start of the project during the architectural definition phase, and towards the end of a project, when the two developments are integrated in a lab environment. That leaves a significant amount of room for each team to have its own interpretation of the architecture and specifications, and for significant differences of understanding to develop.

Under this scenario, the hardware team develops an implementation that follows their interpretation of the architecture, while the software

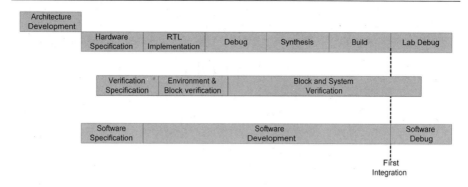

Figure 6-4: Independent Hardware and Software Design Flow

team does the same. Since the software team does not have access to a hardware reference model, each group develops according to their own interpretations. Figure 6-4 shows a classic project timeline where hardware and software development is done independently. There is very little interaction between the two developments until the point where the hardware has been designed and built. This usually means that most, if not all, modifications must be made by software, even when a hardware solution would have been preferable.

Only when the hardware is available in the lab does the software group have access to the hardware environment. At that point, any differences in interpretation show up as bugs. Since the hardware is already developed and usually difficult to change, tradeoffs tend to require software modifications. This is a time-consuming and sometimes frustrating process. The process of developing a single intertwined design flow is often referred to as concurrent engineering. Much has been written about concurrent engineering. Here it will only be mentioned because co-simulation and functional verification are a part of concurrent engineering.

Concurrent Engineering

Co-simulation offers the opportunity for hardware and software to be integrated at an earlier phase of the development cycle. Sometimes

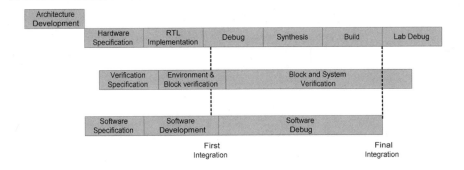

Figure 6-5: Interconnected Hardware and Software Design Flow

called concurrent engineering, the goal is to use the actual RTL as a hardware reference model. By integrating the hardware and software developments early, some projects gain multiple benefits. Figure 6-5 illustrates the potential integration points between hardware and software disciplines. Early and frequent communications allow much earlier and therefore more cost-effective corrections to any potential issues.

Early Communication Between Components

By providing both earlier and more frequent communication between the hardware and software blocks, inconsistencies in the specifications, registers, or functionality can be discovered much earlier. Early detection has several benefits. Earlier discoveries usually mean fewer, simpler, and cheaper modifications. Since the error has not had as much of a chance to propagate, it is likely to show up in fewer places. Similarly, there are usually more and better options for fixing the error, since the hardware has not yet been built. This means that corrections will sometimes be simpler to make.

Since the hardware and software views of an architecture are often somewhat different from each other, co-simulation provides an early way to test that these two different views match each other. Since they are both implemented based on interpretations of the system architecture, they provide a set of checks that both teams implemented in a consistent fashion.

Virtual Lab Environment

Since co-simulation has some similarities to a lab environment, there is the potential for significant re-use between components. In a more traditional environment, the roles of verification code and lab diagnostics are similar, but completely separate. Co-simulation can be used to create an overlap between these roles. While they have different goals, many of the functions are quite similar, and there is a significant potential for code re-use between functional verification, diagnostics, and sometimes the software as well. By sharing an environment, the hardware can be verified at the same time as the diagnostics. By doing this, the lab bring-up process can be significantly shortened, since most pieces will have been tested together before the lab effort starts.

One example of this is the software used to initialize the chip. If there is a single protocol defined for initialization, then it is possible for all three blocks, verification, diagnostics, and software, to share a single piece of software, or at least the same algorithm.

Methods of Co-simulation

While the concept of co-simulation is fairly straightforward, there are many different methods used to tie the hardware and software environments together into a single environment. This section will describe some of the more common methods.

Complete Processor Model

The most straightforward co-simulation method is to use a complete processor model in the simulation, which runs all the existing software. Figure 6-1 provided an illustration of this model. In this case, the full processor model is simulated along with the system. The processor will perform instruction fetches from the memory block, and will execute the code according to those instructions. Everything is contained in the simulation.

To perform co-simulation in this type of model, one must compile the software as one would for a real machine, and load it into the simulated memory blocks. For complete simulation models this is usually straightforward. Complications occur when some blocks are not modeled, such as hard drives or network interfaces. In these cases, software modifications, or additional simulation models may be needed.

Once this is done, the simulation will run the complete environment, including hardware and software. All instruction fetches are cycle accurate in the simulated execution. As long as all I/O devices have a reasonably accurate model, this will provide a very faithful simulation of the architecture.

The major disadvantage with this method is that the simulation can be quite slow. Because the processor is being modeled completely, the size of the simulation may be quite large, and each compiled assembly instruction fetch and decode is simulated, which requires a significant number of cycles. As a result, it is generally impractical to run anything more than some simple drivers in the co-simulation mode.

Bus Functional Model-based Methods

Most other methods of co-simulation use a different interface for the processor. Rather than having a complete processor model in simulation, they use the concept of a bus functional model that is driven by an external processor model. The bus functional model may either be the front end of the processor, or a separate bus model that is responsible for driving and receiving bus operations.

In the complete processor model, the software was executed in a processor that resides in the simulation. While this is very accurate, it is also slow, since the processor tends to be a large device. All other

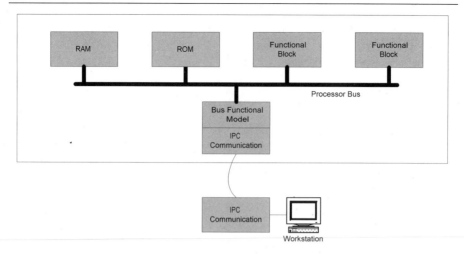

Figure 6-6: Co-simulation with a Bus Functional Model

methods of simulation revolve around executing the software code somewhere else. This may be in an actual external processor that is linked into the simulation, or in a more abstract model of the processor that can run much faster.

In all cases, the hardware simulation must communicate with a processor model that exists somewhere else. In Figure 6-6, the processor model has been replaced with a bus functional model. This is very similar to the transactors, which have been described in Chapters 4 and 5. This bus functional model must then communicate with a processor model, often using some type of interprocess communication (IPC).

Since the processor model is disconnected from the bus functional model of the processor bus, some type of IPC is necessary to connect these two models together. How these two models communicate depends on the type of processor model to be used, and the performance requirements. The possibilities range from a simple socket or pipe to a more dedicated hardware solution such as a PCI bridge.

There are several common processor models: the use of a dedicated hardware processor, such as one might find in an in-circuit emulator;

the use of a model of the processor, often referred to as an instruction set simulator; or compilation of the software onto a different processor altogether. In this last case, the most frequent target of the compile is the same machine that the simulation is running on.

In-circuit Emulation

One common method to provide a high-performance and highly accurate model of the processor is to use an in-circuit emulator (ICE). This provides an actual external processor that is able to run at very high speed, generally much faster than the rest of the simulation. The software, and any applicable operating system, can all be run on the ICE. Many ICEs also provide a powerful debugging environment so that the software operation can be tracked and stepped, and processor registers can be examined.

Generally the software is run on this computer just as it would be run on the intended device. Any access to specific simulated hardware is trapped and sent via the IPC to the simulated hardware. Memory accesses can be made either to the local computer memory at full speed, or can be trapped and sent to the simulator to be run at very low speed.

Since an external processor is almost always a different machine (although one could conceptually run the simulation on the target processor), the IPC is usually done with a socket connection between the external processor and the simulation. Figure 6-7 illustrates software running on an external processor.

There are multiple advantages to using an ICE. Since the processor is real, it provides great accuracy. The simulation can be configured to run highly accurately, so that every cycle communicates between the simulation and the ICE, or it can be configured to run very fast, where only transactions that communicate with the hardware are sent to the simulation. In this case, the instruction fetches are done within the ICE.

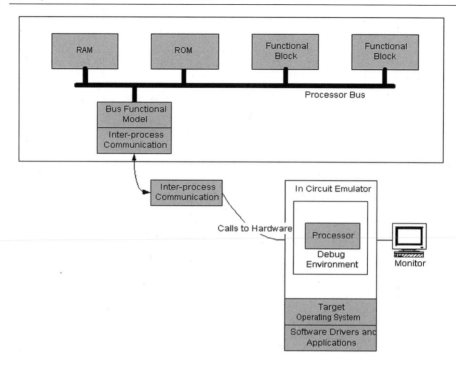

Figure 6-7: Co-simulation with an In-circuit Emulator

The primary disadvantage of an ICE is that a physical setup is required for each simulation. In a larger project, where there are many simulations being run simultaneously, it can be both expensive and time-consuming to manage a reasonable number of parallel simulations.

Instruction Set Simulator

Where an ICE uses an actual processor to connect to the simulation, an instruction set simulator (ISS) avoids this by using a software model of the processor. The processor model in an ISS is generally abstract enough to allow it to run fast, yet accurate enough that it can run the processor assembly code in the same way as an ICE.

An ISS will usually also have a variety of debugging tools that provide control over the program flow, and allow the user to examine the processor registers and other internal state. With the ISS connected to a simulation environment, debugging complex hardware/software interactions is aided by having visibility into both environments.

Connecting the ISS to the simulation is generally quite similar to how an ICE would be connected. As was seen in Figure 6.7, a bus functional model is still needed in the simulation environment. The bus functional model is then connected to the ISS through an IPC protocol. The co-simulation code and potentially the target operating system may both be running inside the ISS.

There are many commercial tools that fit into this model. Most of these tools will provide an ISS that can run the software code, as well as a host of debugging tools to provide a great deal of visibility into both the software and the processor. They will also provide a link from the processor model into the simulator, and many will also provide a bus functional model that can be inserted into the hardware simulation model.

One major advantage of an ISS over in-circuit emulation is the ease with which software models can be replicated. Where an ICE requires physical hardware to be installed and maintained for each simulation, because an ISS is a software tool, it is easier to manage when many parallel simulations are run.

Accuracy vs. Speed in an ISS

Once the ISS is connected up, there are two possible simulation methods that can trade off accuracy and speed of the simulations. At issue is whether the ISS should execute instruction fetches from the hardware simulation memory, or if instruction fetches should be handled within the ISS.

The accurate model will have all the instruction fetches sent to the hardware simulation. This causes the appropriate bus traffic on the processor bus, and the appropriate memory utilization. This method will provide accurate performance and bus utilization information, which may be important if there are bottlenecks in these parts of the system.

The higher-speed model will have all of the software, including instruction fetches, run within the ISS. In this case, the only time that the ISS will communicate with the hardware model is when there the software executes a specific call to a hardware block. All other code remains within the ISS. This mode tends to be faster, since the hardware model is generally two to four orders of magnitude slower than the ISS, depending on the size and complexity of the hardware design.

Usually, an ISS is able to support both modes. Deciding which mode to use will depend on the goals of a particular run. If a great deal of software is to be run, such as booting an operating system, then the speed mode may be required. On the other hand, if performance estimates are needed, or if highly accurate interactions are needed, then having all processor accesses go to hardware is important.

For either of these models, it is important to determine how interrupts are to be handled, and the accuracy of interrupt latency that is required. If an ISS is communicating infrequently with the hardware model, as in the high-speed mode, then the accuracy of the interrupts may be quite low. It also may not fully test the software, since interrupts may only be received at specific times in the software model. Interrupts are just one example of a number of issues that may change how the system behaves if not modeled sufficiently accurately.

Native Co-simulation

One other method avoids using a specific processor model at all. Instead the software and drivers are compiled and run directly on the workstation that is also running the simulation. This is sometimes referred to as native code execution.

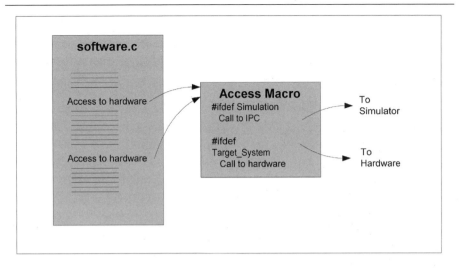

Figure 6-8: Native Co-simulation

Where the previous methods involve compiling the software and loading the resulting assembly code into memory, to be executed by a processor model, in native execution, the software is compiled directly into the workstation, and it relies on the operating system that is running in the workstation.

In native simulation, the software is written so that all accesses to hardware are done through a set of functions, or use a specific data type. The functions have macros defined that will route the call to the IPC in a simulation mode, or call the hardware directly when in the target system. In Figure 6-8, the interconnect between a software block and the access macros is shown.

When specific data types or address ranges are available, then operator overloading can be used to redirect accesses to hardware address ranges from the software environment to the hardware simulator. In either case, this is often one of the easiest methods to connect software code to a hardware simulation.

This method has significant advantages and disadvantages. On the positive side, native co-simulation is usually the easiest method to connect, it is very fast, since the workstation is likely to be much faster than any processor model, and it is low-cost, since there are no additional software or hardware tools to buy.

However, there are also multiple disadvantages that often rule this method out. Possibly the most important is that the software must run on the operating system of the workstation, rather than the target operating system that will be run in the device. If the software relies on specific operating system features, such as timers or system calls, then native co-simulation is likely to fail for that software.

Additionally, the accuracy of this model is low. As with the high-speed mode discussed for an ISS, the native co-simulation model will only communicate with the hardware simulation model when the software calls a hardware function.

Each of these models has various tradeoffs. One difference with native co-simulation is that it is sufficiently simple and low-cost that it is sometimes worth using in addition to other methods. As sections of code become available, it can be easy to test a code fragment this way as soon as it is written. A simple example is a reset sequence. If there is a software module responsible for putting the hardware into a known reset state, that software may very well be able to run in this environment. Similarly, many code fragments that are used to set up or verify the hardware should also be able to run.

When this method is able to work with code fragments, not only can it be used to integrate the hardware and software early, but it can also allow one section of software to serve in verification. If there is an actual piece of software to perform a specific function, then there should be no need to write that function in verification. Alternatively, if the software is not yet available, the verification code should be usable as a section of software once that group is ready to incorporate it.

Acceleration-based Methods

All of the methods described so far use a simulator to model the hardware implementation. While this makes for a fairly simple and highly accurate model, it is also slow. When the goal is only to run some code fragments through the system, this may not be a problem. However, if one wants to run an entire operating system boot sequence, this approach is not likely to be practical.

For reasonably complex systems, most simulators are likely to run at around 100 Hz. This may be higher or lower, by an order of magnitude or two. However, this number must be compared with a likely target frequency of 100 MHz, within an order of magnitude or two. This means that there is an inefficiency of approximately six orders of magnitude. For a software sequence that would require 10 seconds in a real system, this inefficiency would result in a simulated sequence that would require approximately four months of simulation time, which is outside the realm of practical solutions for most applications. This is especially true if a bug is found. One of the most painful parts of very long tests is when a bug is hit near the end of the test. The debug–fix–validate sequence can take a very long time.

As a result, applications that are intended to run significant amounts of code on an RTL implementation model may benefit from using one of many acceleration methods in place of a traditional simulator.

Choices Based on Project Requirements

Choosing an appropriate method for co-simulation in a project depends on matching the requirements of the project with the various methods. There are several issues that need to be considered when selecting a method of co-simulation.

- *Simulation accuracy* – How critical is the cycle-by-cycle accuracy of the hardware/software interface?

- *Simulation speed* – How many cycles need to be run to produce results?

- *Software visibility* – How important are software debugging and control tools?

- *Parallel simulation* – How many simulations need to be run in parallel, and is that practical if hardware is involved in the co-simulation?

The most important requirements are likely to revolve around the software. If there are interactions between the software and the operating system, then it must be run in an environment that supplies the target operating system. If the software is in assembly language or object code, then the co-simulation environment must have either the target processor, or a good model of one. It is important that the chosen method is able to support the existing software with little or no modification.

A second set of considerations revolve around the project requirements. How fast do simulations need to run to be effective? Some methods will provide greater accuracy, but they will be slower. Can one simulation run quickly enough to meet the objectives of the co-simulation environment? Similarly, how many simulations need to be run in parallel? In a large regression environment, there can be a large number of simulations running on a server farm. If the co-simulation environment has a hardware component, such as a processor board, this may not be an acceptable environment.

Finally, the level of control and visibility needed by the software team may be a consideration as well. An ISS, or native co-simulation will both support a debugger that permits visibility and control over the software. Other methods, such as a complete processor model in RTL may provide less visibility into the software.

Summary

Co-simulation can be a powerful tool for integrating the hardware and software blocks of a full system architecture. With the early integration of the two disciplines, there can be substantial benefit in verifying that the architectural components work together. Since software and verification often play similar roles in a system, there is the ability to share algorithms and code between the two teams, and provide much more interaction between the two teams at an early stage. Scheduling between the teams is a crucial component of co-simulation. There are generally many hand-off points where one team must deliver code to the other. Without integration between the hardware and software schedules, co-simulation is unlikely to work successfully.

While the benefits of co-simulation can be significant to the verification task, care must be taken to ensure that the methods and tools used are sufficient to meet the goals. Co-simulation can be computation-intensive, with significant setup efforts. If this is not well planned, the benefits may never materialize, or not be worth the effort. Done well, it may provide important and timely benefit to the overall project.

Measuring Verification Quality

Key Objectives

- Understand the issues surrounding verification quality

- Discuss the methods used to estimate verification quality

- Compare the strengths and weaknesses of the various methods

- Examine real-world considerations

It is very important to be able to estimate the quality of functional verification in order to understand the likelihood of success when a design is implemented in a real system and to determine when functional verification is complete. It can rarely be stated with any certainty that there are no functional issues left in an implementation of an architecture. However, there are methods available to measure functional coverage, and to have some degree of confidence in the quality of the architecture and implementation. Since it is never known if all bugs have been found, or how many bugs are actually in a system, the goal is to achieve as accurate an estimate as possible of verification quality.

In measuring the quality, it is important to remember that the goal is to verify how well the implementation matches the architecture. This is not the same as measuring how well an implementation has been tested.

Estimating Quality

At the point in a project where an RTL design must be implemented in hardware, if the verification quality is acceptable, then that hardware will function correctly when it is installed in a system. If there is a functional bug in the hardware that prevents the system from working correctly, then that is generally due to insufficient functional verification. Since the decision to build an implementation is so critical, it is important to understand the quality level of the verification test suite that has been used to check the design.

No Proof Is Available

There is no way to know that all functional bugs have been removed, or how many bugs are still in existence for any design of non-trivial complexity. All that any tool can provide is an estimate of quality based on some assumptions. To relate this issue to a well-known example, consider the case of spelling errors in a text document. A spell-checker will ensure that all words are legal, but it will not be able to recognize errors that result in other legal words, for example, it will accept either "form" or "from." Once a spell-check is complete, the quality level of the documents can be further improved using a grammatical check tool, and by proofreading it. All of these methods are effective, and they raise the estimated quality level of the document, but it is still possible for an error to be missed by all of these checks. There is no guarantee that all the errors have all been found. The same is true for any architectural implementation. There are many methods to find errors and estimate the quality level of a design, but there is no combination of methods that prove that all errors have been found.

Approximation of Verification Quality Is Required

Because there is no proof, there is no way to determine if most or all bugs have been found. The goal is to provide an accurate estimate of the quality of a verification suite. In providing accurate estimates, it is important to note that bugs are found in all areas of a project, not just the RTL design. They will be found in the architecture, design specifications, RTL designs, and software code. In a system, hardware that faithfully implements an architectural bug will still fail.

So, the issue that projects face is determining when functional coverage is complete, and the implementation can be built. An accurate estimate of verification quality is necessary to make an intelligent risk/reward tradeoff when determining if the verification is complete. If the implementation is to be built in an ASIC, then the cost of making a wrong decision can be very high. Time is usually the biggest cost, since it represents both lost opportunity for the product to be sold on the market and the cost associated with keeping an entire project team working on that same project. In many cases, the cost of actually rebuilding the ASIC is less than the accumulated time costs required to debug, re-verify, and re-implement the design, and then convert the design into a new ASIC. This must be traded off against the cost of finding bugs in simulation. As fewer bugs are found, the time between bugs in simulation increases. As this time becomes larger, a point will be crossed where it becomes less expensive to build hardware and find bugs in a real system than to continue with only simulations.

Even with a more easily modified implementation, going from simulation to hardware too soon can carry significant costs. While field programmable gate arrays (FPGAs) are often smaller and cheaper to fix than ASICs, tracking down issues is still much harder in the physical implementation than in a simulation environment, due to restricted visibility and difficulty of making modifications. Changes that are beyond a small local change can be quite painful, since they may require pin and board modifications. Of course, all of that is simple compared to the cost and time involved in making a modification to an ASIC.

Assumptions Used in Determining Verification Quality

In providing estimates of functional coverage, most methods rely on assumptions that are often implicit. It is important to understand the underlying assumptions of each method, since this is where the measurements and reality can diverge. Each method will provide a very specific measurement. For example, a code-coverage tool may report 92.6% coverage. This sounds very accurate, and in fact it is accurate in terms of what it has measured. It is easy to jump to the conclusion that this reflects an accurate measure of quality. There is an underlying set of assumptions that relate the particular measurement to the verification quality level. Those assumptions need to be understood in order to determine how closely a particular measurement relates to - the actual verification quality.

One reason that understanding the assumptions has become important is due to an industry trend to provide tools that improve verification quality metrics. What the tools do is exactly that: improve the metrics. It is implied that an improved metric means improved verification quality. To determine the accuracy of that relationship, it is important to understand what assumptions are being made in relating the particular measurement to an estimate of verification quality.

Overview of Methods

There are several common methods used to estimate the quality of functional verification. They use different approaches to measure an aspect of verification quality. Some are based on measuring aspects of a simulation and are tool based, others are more project methodology based. Because these are different approaches, the best overall measurements are often a combination of several techniques.

In looking at these methods, it is important to remember what each of them is measuring, and how that measurement relates to actual architectural functional coverage. Five common methods are:

- *Code coverage* – A simulation-based method that in its simplest form counts which lines of RTL have been exercised when a verification suite is run.

- *Functional coverage* – Checks that the known valid states of a particular block of code have been exercised.

- *Fault insertion coverage* – Inserts a bug and see if the test suite catches it. Uses a set of bugs to measure test suite quality. This can be either complete fault coverage, where all possible bugs are inserted, or statistical fault coverage, where a random population of faults are inserted.

- *Bug tracking* – Tracks all found bugs, and correlates them to the RTL. Looks for discrepancies between bug rates and code quantity and complexity.

- *Design and test plan reviews* – Compares the verification test plan with the architectural and implementation specifications. Looks to see that all components of the specification have a corresponding test.

Figure 7-1 and the subsequent listings illustrate the differences among the various methods. In Figure 7-1, a highly simplistic example shows a state diagram of a very simple up-down counter. While it is unlikely anyone would create an up-down counter that looks like this, it is useful for illustration purposes.

The up-down counter is implemented in a state-counter module shown in Listing 7-1. Again, it is unlikely that an up-down counter would be implemented as a state machine under normal circumstances, but this makes for an easy-to-understand example of a block of control logic. Note that there are several coding styles in this example that are not recommended. They are being used to illustrate some of the issues around coverage analysis.

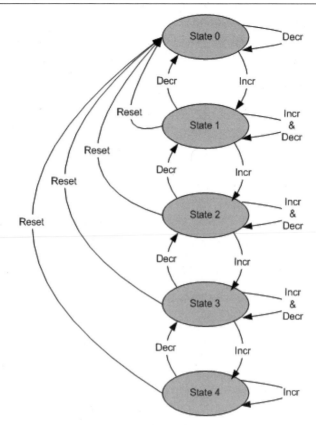

Figure 7-1: A State-machine Counter

A verification test is shown in Listing 7-2. This is a simple sequence that attempts to test the state counter by resetting it and counting all the way up and all the way back down. Some observation will show that this is not a complete test. These examples show the differences among the methods of verification quality.

Code Coverage

The concept behind code coverage is to measure which parts of the implementation have been exercised by the functional test. To do this, various software tools are used in conjunction with the simulation to monitor the implementation and track the RTL during a simulation run. There are various kinds of code coverage.

Listing 7-1: State_counter Code

```
1   module state_counter(clock,
2                        reset,
3                        incr,
4                        decr,
5                        state);
6
7       input           clock;
8       input           incr;
9       input           decr;
10      input           reset;
11
12      output [3:0]     state;
13
14      reg [3:0]        state;
15      wire             incr;
16      wire             decr;
17      wire             reset;
18      wire             clock;
19
20      initial begin
21          state = 2;   // start is some non-X state for simple example
22      end
23
24      always @(posedge clock) begin
25          case (state)
26          0: begin
27              if (incr == 1) state <= 1;
28          end
29          1: begin
30              if        ((incr == 1) && (decr == 0)) state <= 2;
31              else if ((decr == 1) || (reset == 1)) state <= 0;
32          end
33          2: begin
34              if (reset == 1) state <= 0;
35              else if ((incr == 1) && (decr == 0)) state <= 3;
36              else if ((incr == 0) && (decr == 1)) state <= 1;
37          end
38          3: begin
39              if (reset == 1) state <= 0;
40              else if ((incr == 1) && (decr == 0)) state <= 4;
41              else if ((incr == 0) && (decr == 1)) state <= 2;
42          end
43          4: begin
44              if (reset == 1) state <= 0;
45              else if ((incr == 0) && (decr == 1)) state <= 3;
46          end
47          endcase // case(state)
48      end // always @ (posedge clock)
49  endmodule // state_counter
```

Toggle Coverage

The simplest coverage measurement is referred to as toggle coverage.
This method tracks the logical value of each net and register in an
implementation. The coverage tool reports any net that does not

Listing 7-2: State_test Code

```
1  module state_test (clock,
2                     state,
3                     reset,
4                     incr,
5                     decr);
6
7      input          clock;
8      input [3:0]    state;
9
10     output         reset;
11     output         incr;
12     output         decr;
13
14     wire [3:0]     state;
15     reg            incr;
16     reg            decr;
17     reg            reset;
18
19     integer        i;
20
21     initial begin
22         incr <= 0;
23         decr <= 0;
24         reset <= 1;
25         @(posedge clock);
26         reset <= 0;
27         @(posedge clock);
28
29         for (i = 0; i < 5; i = i+1) begin
30             #1;
31             $display ("State: %d, i: %d", state, i);
32             if (state != i) begin
33                 $display("Error on up_count");
34             end
35             incr <= 1;
36             @(posedge clock) ;
37         end
38         incr <= 0;
39         for (i = 4; i > 0; i = i-1) begin
40             #1;
41             $display ("State: %d, i: %d", state, i);
42             if (state != i) begin
43                 $display("Error on down_count");
44             end
45             decr <= 1;
46             @(posedge clock) ;
47         end
48         decr <= 0;
49         $finish;
50     end // initial begin
51 endmodule // state_test
```

toggle during the test. Some tools require only that a net takes on both a low and a high value in order to be counted as a covered net. Others require a transition to both states. Requiring transitions is a slightly more rigorous method, since nets that start in one state at time 0 are not counted as partially covered.

The strengths of toggle coverage are that it is easy to implement, and any net that has not toggled has certainly not been tested, since it has not been exercised. The major weakness is that toggle coverage has probably the highest percentage of false positives of any of the coverage tests. The reason is that a great deal of logic will start toggling simply by injecting a clock. Counters will start counting, and a significant percentage of the nets in a design may switch states without any external test activity at all. It is not unusual to achieve a 30% coverage measurement simply by toggling clocks and asserting a reset. While the toggle test has the most false negatives, all coverage-based tests are susceptible to this. Just because a net has toggled does not mean that the result of the toggle has been tested and verified.

In the example test, there are only a few registers and wires that can be examined by toggle coverage. This would include the clock, reset, increment, decrement, and state signals. All of them will have been at both values multiple times, so the toggle coverage should be reported as complete. The state will have taken on only five legal values out of the possible 16, so some types of value coverage might report that the register did not take on all possible values. In this example, it did take on all legal values.

Line Coverage/Block Coverage

Line coverage, which may also be referred to as block coverage when used on blocks of RTL, measures that each line of RTL has been exercised during a simulation run. In this case, the coverage tool monitors the progress of the simulation and tracks which paths of RTL code the simulation is executing. At the end of the simulation, any lines that have not been executed are marked as uncovered.

Line coverage is often more thorough than toggle coverage, since it actually measures that code was executed. However, it too has a significant number of false positives. As before, simply injecting a clock will cause some not-insignificant number of lines to be executed before a test even has a chance to run. Coding style is also important to the effectiveness of line coverage. Too much logic on one line of code results in an overly optimistic coverage report. For example, putting both an "if" statement and its corresponding "else" statement on the same line will result in both branches being marked as covered if either of them are executed. Another example of this can be seen on line 31 of Listing 7-1. In this case, both the reset and the decrement conditions are put on a single line.

Running the Example with Line Coverage

Running the example test with line coverage would report the first weakness of the test. Each state has a separate line for reset. The test issues only a single reset once at the beginning of the simulation. So, this metric would indicate that the test is not sufficient, and that it should issue a reset from each state of the counter. Listing 7-3 highlights the uncovered lines from the example test above.

While this is clearly a contrived example, there is still a critical point that is being shown. The line coverage indicates a weakness of the test with respect to the RTL design. This may not have been specified in the architectural specification. It is important to remember that the primary purpose of functional verification is to verify that the implementation matches the architecture. Modifying or adding tests based only on coverage metrics may not be working towards that goal.

Note that by writing code across more lines, the quality of line coverage can be improved, although the code will be more verbose.

Listing 7-3: Line Coverage Results

```
1   module state_counter(clock,
2                        reset,
3                        incr,
4                        decr,
5                        state);
6
7       input           clock;
8       input           incr;
9       input           decr;
10      input           reset;
11
12      output [3:0]    state;
13
14      reg [3:0]       state;
15      wire            incr;
16      wire            decr;
17      wire            reset;
18      wire            clock;
19
20      initial begin
21         state = 2;   // start is some non-X state for simple example
22      end
23
24      always @(posedge clock) begin
25         case (state)
26            0: begin
27               if (incr == 1) state <= 1;
28            end
29            1: begin
30               if      ((incr == 1) && (decr == 0)) state <= 2;
31               else if ((decr == 1) || (reset == 1)) state <= 0;
32            end
33            2: begin
34               if (reset == 1) state <= 0;
35               else if ((incr == 1) && (decr == 0)) state <= 3;
36               else if ((incr == 0) && (decr == 1)) state <= 1;
37            end
38            3: begin
39               if (reset == 1) state <= 0;
40               else if ((incr == 1) && (decr == 0)) state <= 4;
41               else if ((incr == 0) && (decr == 1)) state <= 2;
42            end
43            4: begin
44               if (reset == 1) state <= 0;
45               else if ((incr == 0) && (decr == 1)) state <= 3;
46            end
47         endcase // case(state)
48      end // always @ (posedge clock)
49   endmodule // state_counter
```

As an example, the if statement on line 31 of Listing 7-3 could be written as multiple nested if statements. This would provide line coverage for each individual portion of the if statement. Alternatively, a branch coverage tool will perform a similar function.

Branch/Path Coverage

Branch or path coverage is quite similar to line coverage. The major difference is that this will actually analyze the RTL and determine if each branch of each line of code has been taken. As before, there will be a significant number of false positives. The major difference is that there are far more coverage tests being done in the case of branch coverage. A reasonably complex code snippet will have many more branches than lines of code.

Running the Example with Branch Coverage

Branch coverage looks at many more conditions, and our simple test illustrates that nicely. The results of a branch coverage test are shown in Listing 7-4. Note that the shaded lines without boxes indicate the lines that were found in branch coverage that were not uncovered in the line coverage example.

In this case, it shows that while we did execute line 31 on our way from state 1 to state 0, we only exercised that path due to the decrement condition, and not due to the reset condition.

Functional Coverage

This method measures verification quality by using knowledge about the operation of a block of code. Note that the internal knowledge of a block is often known only to the designer and not the verification engineer. As a result it is usually the designer who must list all known states or operations of any particular block. The functional coverage can then check that each entry in the list has occurred sometime during the verification test suite. Any entry in the

Listing 7-4: Branch Coverage Results

```
1   module state_counter(clock,
2                        reset,
3                        incr,
4                        decr,
5                        state);
6
7       input           clock;
8       input           incr;
9       input           decr;
10      input           reset;
11
12      output [3:0]    state;
13
14      reg [3:0]       state;
15      wire            incr;
16      wire            decr;
17      wire            reset;
18      wire            clock;
19
20      initial begin
21          state = 2;   // start is some non-X state for simple example
22      end
23
24      always @(posedge clock) begin
25          case (state)
26              0: begin
27                  if (incr == 1) state <= 1;
28              end
29              1: begin
30                  if      ((incr == 1) && (decr == 0)) state <= 2;
31                  else if ((decr == 1) || (reset == 1)) state <= 0;
32              end
33              2: begin
34                  if (reset == 1) state <= 0;
35                  else if ((incr == 1) && (decr == 0)) state <= 3;
36                  else if ((incr == 0) && (decr == 1)) state <= 1;
37              end
38              3: begin
39                  if (reset == 1) state <= 0;
40                  else if ((incr == 1) && (decr == 0)) state <= 4;
41                  else if ((incr == 0) && (decr == 1)) state <= 2;
42              end
43              4: begin
44                  if (reset == 1) state <= 0;
45                  else if ((incr == 0) && (decr == 1)) state <= 3;
46              end
47          endcase // case(state)
48      end // always @ (posedge clock)
49  endmodule // state_counter
```

list that has not been exercised sometime during the test is reported as uncovered.

The simple example above might list all of the states in the counter state machine, but this would be too simplistic an example to have much meaning. In real projects, this method is more likely to be used for more complex blocks. For example, functional coverage of a PCI bus might have a list of all legal PCI operations, plus require back-to-back reads and writes, as well as any interesting error conditions. If the system under test is expected to handle parity errors, then the functional coverage list for the PCI bus would be expected to have a set of parity error conditions included.

A major advantage of functional coverage is that it provides a more system-level view of coverage. By reporting on coverage of functional blocks, it is easier to understand the implications of any non-covered events.

Cross-product Functional Coverage
In more complex systems, there will be multiple places where functional coverage is tracked. Functional coverage is used to track each resource individually. A cross-product analysis can be used to examine the coverage of both resources combined. For example, in a system that provides a bridge between two PCI buses, functional coverage may track the states of each bus. A cross-product analysis may be used to show if all possible combination of states were covered.

As a larger number of buses, state machines, and interfaces are monitored via functional coverage, a complete cross-product analysis becomes impractical. It is important to determine which resources may be related, and provide cross-product coverage analysis on those resources.

Limits of Functional Coverage

As with line coverage, simply exercising a particular operation does not mean that the operation worked correctly, or that the value was checked by the test suite. In the PCI example, if a parity error is driven onto the bus on a write operation, but the result of the operation is never checked, perhaps by reading the operation back, then no error would be detected. The functional coverage report would show that a parity condition had been exercised. While that is true, the result was never checked, so there is no knowledge of whether the parity condition worked or not.

Functional coverage also differs from other coverage techniques because it relies on statements being made to monitor the coverage of any particular device. Unlike other coverage methods, if a statement is not made, then coverage is not monitored. As a result, functional coverage can be useful for monitoring expected functions, but it will not help in the detection of unexpected problems. Unfortunately, most complex designs have a significant number of functional bugs that were due to unexpected interactions between blocks. Functional coverage is often not the ideal method for determining the likelihood that all interaction problems have been tested. Other coverage metrics, which do not rely on an engineer to determine the measurements, are often better at detecting uncovered areas of a design.

Implied Logic

Note that none of the coverage tools indicates whether implied functions have been covered. This is logic that is part of the design, but for which there are no lines of code. For this reason, as well as for clarity of the design, it is usually a good idea to avoid implied logic.

Of course, the state-counter in Listing 7-1 would not have been complete without an example of implied logic. In each of the states 1, 2, 3, and 4, there is an if statement that does not have a default else statement. In each of these states, if the increment and decrement signals are both asserted simultaneously, then the state machine will remain in the current state. This behavior is not specified in the RTL. Rather, it is due to the fact that the state registers have not been changed that this behavior will occur.

Because there is no code to implement this functionality, the coverage tools have nothing to measure. There will be no indication if this situation has been tested or not. In order to avoid this, an else condition should be added to every if condition, and a default case should be added to case statements.

Fault Insertion Coverage

A somewhat different model for functional verification is fault insertion coverage. This is an older technique that originates from hardware-manufacturing test suites. It was developed to check the quality of manufacturing test suites. This method is rarely used today, but it is useful to discuss, since it illustrates a key weakness of the code coverage tools.

The premise is to artificially introduce a single fault into the implementation and to run the entire verification test suite to see if the fault is detected. If the test suite detects the error, then that particular fault is covered by the suite. This process is repeated for every possible fault.

A critical difference between fault coverage and the previous coverage techniques is that this metric checks that the fault is actually detected by the verification test code. Simply causing the error is not sufficient. This test will pass only if an introduced error is detected by the verification test suite.

When an inserted fault is detected by the tests, it means that the tests created the stimulus necessary to exercise the particular fault, and that the results were analyzed sufficiently to catch the change in outputs. That is a good indication of the quality of the tests, but it should be remembered that it still does not ensure that there are no bugs in that particular piece of the design.

Because any reasonably sized design has an enormous number of potential faults, and a verification test suite will often run for multiple CPU-days, a complete fault-insertion test is generally not practical despite its more rigorous measurement methods. However, there are some types of applications, often in mission-critical or fault-tolerant systems, where a more thorough coverage metric is useful.

It is possible to obtain fault coverage metrics for a verification suite without performing a full fault-insertion test of every possible fault. This may be done with some statistical analysis. Since the complete set of faults is generally quite large, it can be considered to be an infinite population. With a random sample of several hundred faults, assuming an infinite population of potential faults, it is possible to predict the actual fault coverage metric within a fairly small range with a 99% confidence level. Most statistics textbooks will cover the large-sample confidence interval concept that is used to make this type of confidence statement.

Running the Example with Fault Insertion Coverage

Running a fault-grade analysis on the example code would reveal still more code that the example verification test did not check. Notice in Listing 7-5 that, in this case, line 31, the heavily highlighted line, was exercised, but never checked. The test caused a transition of the state machine from state 1 to state 0. However, there was nothing in the test that checked that the state machine ever arrived at state 0. If the code were incorrect, the test would not have caught it.

This occurred because of a weakness in the test. In Listing 7-2 on line 45, the decrement condition has been exercised. However, the check condition was done earlier, on line 40. On the final time through the loop, there is no final check condition.

Listing 7-5: Fault-grade Results

```
1   module state_counter(clock,
2                          reset,
3                          incr,
4                          decr,
5                          state);
6
7   input           clock;
8   input           incr;
9   input           decr;
10  input           reset;
11
12  output [3:0]    state;
13
14  reg [3:0]       state;
15  wire            incr;
16  wire            decr;
17  wire            reset;
18  wire            clock;
19
20  initial begin
21      state = 2;   // start is some non-X state for simple example
22  end
23
24  always @(posedge clock) begin
25      case (state)
26          0: begin
27              if (incr == 1) state <= 1;
28          end
29          1: begin
30              if      ((incr == 1) && (decr == 0)) state <= 2;
31              else if ((decr == 1) || (reset == 1)) state <= 0;
32          end
33          2: begin
34              if (reset == 1) state <= 0;
35              else if ((incr == 1) && (decr == 0)) state <= 3;
36              else if ((incr == 0) && (decr == 1)) state <= 1;
37          end
38          3: begin
39              if (reset == 1) state <= 0;
40              else if ((incr == 1) && (decr == 0)) state <= 4;
41              else if ((incr == 0) && (decr == 1)) state <= 2;
42          end
43          4: begin
44              if (reset == 1) state <= 0;
45              else if ((incr == 0) && (decr == 1)) state <= 3;
46          end
47      endcase // case(state)
48  end // always @ (posedge clock)
49 endmodule // state_counter
```

While this is a simple, and clearly contrived example, in more complex systems this becomes very critical. A test that is verifying some very specific functionality may well trigger conditions that it does not care about or check. Coverage tools alone will not provide any indication that sections of logic have not been verified. This is when the coverage tool gives a false positive report, and it is a problem with poor tests. A poor test may create a lot of stimulus and cause a great deal of activity inside the design while actually testing very little. An extreme example of this would be a test that injects a large number of transactions, but never checks a single result. This test may show good coverage results even though it never checked anything. A good test will also create activity, but will check for correct behavior in the areas where it is creating activity.

This illustrates one issue of the normal coverage tools. Toggle, line, and branch coverage are all negative tools. They are able to report that something is missing in the tests. However, the lack of such a report does not indicate that the tests are complete. It is necessary but not sufficient to check that a line of code has been covered. For a verification test to be effective, the consequences of each covered line of code must also be verified to ensure that the code took the right action.

Bug Tracking

Despite the decidedly low-technology approach, bug tracking can provide a wealth of information about a project. Bug tracking is simply recording all bugs that are found by engineers throughout the project, not just in RTL implementation, but also in the documentation and software. A bug in any part of the system may result in a failure of the device. A hardware design that accurately reproduces a bug in the documentation will result in a hardware failure.

A good method for recording all information about the bug is required. At a minimum, this includes when the bug was found, when it was resolved, where it was found, and in the case of RTL code, the file and logical block that was involved. The purpose of this information is to track the distribution of detected errors. There are commercial tools specifically developed for this purpose, some of which can help in the analysis of data information.

To get accurate bug information over the course of a project, a couple of things must be kept in mind. The first is that the bug tracking must begin early in the project. As soon as any part of a project is released to the group, all bugs associated with that project must be tracked. The second is that all bugs must be reported, independent of why or where they were found. An incomplete record of bugs will be an indication of problems that may not exist. This can happen because it will seem that too few bugs were found in an area, when the actual problem is that the bugs were not reported.

Finally, it is important to ensure that any bugs that were found are tracked and checked for the duration of the project. It is not uncommon for a bug to be re-introduced into the design. This may happen when someone forgets why a problem occurred and recreates some code that had been fixed, or when an old code fragment is re-used. In any case, it can be quite frustrating to debug a single bug multiple times. By tracking and continuously checking for all bugs that have been found, this problem can be avoided.

Bug Detection Rate

As a project progresses, there are several interesting statistical measures that can be monitored. The most obvious one is the rate of bugs being found. In many projects there is a natural curve, where early in the project, bugs are hard to find because the hardware is not yet fully functional. As the hardware improves, the bug rate increases for a while. Then, as the hardware becomes more stable, the bug rate decreases, and much more work is required to find each additional bug.

The bug detection rate can provide some indication of the effectiveness of the verification methodology. If the rate does not increase quickly, or if it flattens out, this may be an indication that there is a problem in verification.

Bug Distribution

The distribution of bugs across the various blocks is also worth tracking. Are some blocks receiving more bug reports than might be expected, based on their complexity? Even more important is if some blocks are not receiving sufficient error reports. This may be an indication of a problem with the verification test plan. Unexpected differences in bug distribution are another early indication that there may be problems in the verification process.

With a good bug tracking system and careful project management, this type of analysis is not very time-consuming. Many bug tracking systems provide graphing features to make bug report data easily accessible. Most important, this information is available fairly early into a project. That means that it is possible to detect and correct potential problems fairly early into the project cycle, while there is still time to correct the problem without affecting project schedules.

Design and Test Plan Reviews

The most straightforward quality metric of them all is the verification design review. This is the single lowest-cost measure of completeness, and it alone is able to correlate the verification tests with the architectural intent of a system.

A verification review should be done at least twice during a project. The first time is near the start of the verification effort. After writing a verification plan that specifies all the required tests, a review is done. This ensures that each test will accomplish its goals, that the tests cover what they were intended to cover, and that all parts of the system architecture have been tested.

For a review to be effective, it must examine and compare the architectural features with the verification test plan with the appropriate people. The goal of a review is to see that all of the architectural features are checked in the tests, and that the tests will be effective in verifying the implementation of the architecture. To be effective, this review must examine the flow of each test and examine how it will exercise the architecture, and how the results are to be checked. While not a great deal of fun, this is usually a valuable exercise.

A second review is often done near the end of the project. At this point, many changes are likely to have been made in the tests, in the architecture, and in the design. The second review ensures that the modified verification plan has not missed any parts of the architecture, that the new tests are effective, and that there are no holes in the verification strategy. This review is generally used to determine if the verification task is complete. If it is found that there are several major components that need to be modified, another review after the changes have been made is generally called for. While the reviews are rarely enjoyable, they provide a highly effective and reasonably low-cost mechanism to check that the verification effort is complete.

Runtime Metrics

There are a few other simple metrics that can be looked at. One of the simplest is to measure the simulation runtime and the number of cycles that have been executed in the functional verification process. This is basically a sanity check. If a one million gate design has a few thousand cycles of functional verification, it is a strong indication that something is wrong with the verification. Most verification suites require a number of cycles per gate in order to functionally verify each gate. As the functional verification techniques become more structured and abstract, that number increases. By looking at the number of cycles per gate, and comparing them with previous projects, one can see if the number is too low.

As with other metrics, runtime measurements are a negative tool. They can indicate that there is a problem with the design, but a good runtime metric does not provide any indication of verification quality.

Comparison of Methods

One thing that all of the methods of verification quality metrics have in common is that none of them measure the actual quality of a verification suite. They all provide metrics that are related to verification quality, and it is up to the user to use those metrics to estimate the actual quality of verification code.

The Practical Implications of Coverage Tools

The code coverage tools are one of the more popular methods used to estimate verification quality. It is fairly straightforward to obtain a metric, and there are many commercial tools available.

Analysis of Coverage Data Is Expensive

What is often overlooked is the cost of analyzing coverage data. A coverage tool will provide a great deal of data. Understanding what that data means will take significant time and effort. There are several reasons that a particular line of code may not be covered.

Dead Logic

In the course of implementing a design, there will be times when code was implemented, then forgotten as another path was taken to complete the implementation. This results in code that is left over from an earlier thought sequence. Part of the job of verification is to make sure that any dead code does not disrupt the functionality of the implementation. Coverage analysis may uncover dead code that does not detract from the functionality of the implementation, but is reported as uncovered logic.

Redundant Paths

This is similar to dead logic, but the excess logic is actually connected up. When this is done unintentionally, it may also be reported as uncovered logic, although it is not useful to the functionality of the implementation.

In some cases, redundant paths are necessary. This may occur in fault-tolerant designs, or where glitch prevention is required. In these cases, special attention needs to be paid to this type of logic.

Special Logic

There are many lines of code that are necessary for the simulation to function, but that coverage analysis tools may not recognize. These include clock signals, power lines, resistive pull-ups and pull-downs, and other such lines.

Uncovered Logic

Finally, a coverage tool will also report lines of code that have not been covered but should have been. If a coverage tool reports that a particular line of code has not been exercised, then there is clearly some logic in the implementation that has not been exercised.

In a complex system, understanding why a particular line of code was not covered may be a difficult task. The relationship of one particular line of code to the operation of that block is often understood only by the original designer. Often, that designer has also forgotten the operation of that block.

Understanding why the verification test suite did not cover a particular line may be important for ensuring that a particular function is not just covered but also checked for correct operation.

Coverage Enhancement May Not Improve Quality

The trend towards coverage enhancement attempts to solve the problem of needing to analyze coverage results. Several verification tools will examine RTL coverage metrics and attempt to augment a verification test suite so that most or all lines of code are covered automatically. This will result in higher coverage metrics. Unfortunately, there is no guarantee that this will improve the quality of the verification test suite. This approach will increase the coverage, and add appropriate stimulus, but cannot provide any assurance that the results of the stimulus match the intent of the architecture.

RTL Code May Not Be Following System Architecture

At any point in a design process, a designer may misinterpret an architectural specification, or due to any number of reasons add some RTL code that is not consistent with the system architecture. This is common, and the purpose of functional verification is to find these types of problems.

However, functional verification should be based on specifications, not RTL code. This is one of the ways that specification bugs are found. If there are lines of code that are outside the specification, then it is unlikely that functional verification will catch the code, assuming that it doesn't interfere with normal operation.

One way of finding undocumented implementation code is with code coverage. This requires examining the uncovered code to determine its function. By simply adding test stimulus to cause code to be covered, any discrepancies between architecture and implementation are not likely to be found through this method.

Test Suite May Not Be Checking Results

Most tests are designed to drive some type of traffic into a system under test, and then to examine the results of that particular traffic to see if the behavior worked as expected. A second issue with coverage enhancement is that it will generate additional traffic easily enough, but it does nothing to verify the results beyond whatever checking was already done. That means that the functions that were not covered may well be covered from a stimulus viewpoint, but there may well be nothing checking that the behavior of the uncovered code is correct.

For example, if a coverage enhancement tool discovers that an error condition has not occurred, it could well generate the required stimulus to create an error. If a test was not written to verify the error condition, then there is nothing to ensure that the error handling was correct. At this point the coverage tool would indicate that error handling has been tested, when it may well not have been.

When either of these cases is true, the coverage enhancement would provide a false sense of security in the quality of the verification code. That could be a costly mistake if it results in a non-functional design.

None of this implies that coverage tools are not useful. Used correctly, and with an understanding of what is actually being measured, coverage tools provide a powerful metric for estimating the quality of verification code.

The Basic Methods Are Most Cost-effective

While automated coverage tools are becoming more popular, they do take significant time and resources. The basic methods of verification reviews and bug tracking are much simpler, and less quantitative in their results. However, they are also much less expensive, in terms of engineering time, to implement. Given the large difference in engineering time, a good set of verification reviews will often provide the single most cost-effective verification quality metric, with bug tracking close behind. Using coverage tools without appropriate tracking and reviews is unlikely to be effective.

A Combination of Methods Provides Best Understanding

Because many of these methods work on different principles, a combination of methods will be effective in providing a good understanding of verification quality. It is this understanding that is critical to determining where and how a verification test suite needs to be improved, and when an implementation accurately reflects the architectural intent of the system.

Real-World Considerations

It is important to understand both the risks and costs of a functional verification project early on. Defining these early in a project will permit rational decisions to be made about which methods to use and the likely schedule impact. Keeping these tradeoffs in mind as a project progresses can be difficult. As a result, it is often helpful to review the early criteria when a tape-out decision needs to be made.

When looking at automated tools, it is important to understand the schedule impact of a detailed analysis. This is not a trivial task, since it requires a very specific knowledge set. Simply using automated tools to improve the coverage metrics may not be improving the actual verification test suite quality.

When new bugs are not being found quickly, there are two possible explanations. Either there are very few bugs to find, or the methods being used are ineffective. Unless all other indications point to the former, the problem will generally be with the latter. That means that a review of the verification approach is necessary.

Reviews are never fun, but they should be low cost, both in terms of tool and designer time. This is also the only method that provides some comparison between the architecture and the implementation. This method should not be ignored.

It isn't always possible to find and remove all bugs. In some larger programs, as the test suite nears completion, the cost per bug in simulation may exceed the cost of implementing real hardware to find the next bugs. Understanding of the bug rates and burn rate of a project, along with budgetary and schedule requirements are important.

Summary

The quality of functional verification is what will determine the likelihood of success when a design is synthesized into a real implementation. As a result, it is important to be able to estimate the quality of functional verification. Weaknesses in coverage may indicate that additional functional tests are required in order to improve the overall quality of functional coverage.

The decision to convert the synthesized design to a physical device can be a costly one. Accurate information on the functional coverage is necessary to provide a reasonable tradeoff between risk and quality.

In measuring the quality of functional verification, it is important to remember that the goal is to verify how well the implementation matches the architecture. Most coverage tools and metrics measure how well an implementation has been tested, which is not the same thing. Testing an implementation is necessary, but the tests must also check that the implementation matches the specifications. That can only be done with more manual processes such as test reviews.

8

The Verification Plan

Key Objectives

- Goals of a verification plan

- Phases of creating a verification plan

A verification plan is a very broad set of documents that provides the goals, components, and details of the verification effort. The plan is a critical part of a successful project. Just as with architectural and hardware specifications, the verification plan defines the goals of the verification team, and provides a vehicle to ensure that everybody, not just the verification team, agrees and adheres to them. During the project execution phase, the plan serves as a reminder of the goals and requirements of the verification project, and is instrumental in determining when the verification effort is complete.

The verification plan must deal with a wide range of issues, from the high-level goals and architectural features that need to be tested, to the low-level identification and documentation of tests and components. As a result, this is generally a significant living document, written by the verification team, that is approximately the equivalent of a functional specification for the design team.

Goals of the Verification Plan

The verification plan is often the central organizational point for the verification of a project. In many cases, the verification effort is also the point of integration between blocks of hardware, third-party intellectual property, and software components. As a result, the verification plan can be an essential element for planning and organizing a substantial portion of the entire project flow.

There are many goals for a verification plan; they fall into a few main areas. These include the documentation of the architecture of the verification environment, a mechanism to ensure that the verification project will cover the architectural and design components of the system, guidelines to determine when the design should move from simulation to implementation, and a roadmap to provide guidance and scheduling for a verification team.

As the verification components that will be required for the project are listed, it is possible to identify potential shared components and applicable existing or commercial intellectual property that can be evaluated or integrated. The verification plan can define an environment structured so that components built in the early phases of the project will fit into later environments as the verification migrates from block to system level, while maximizing component re-use.

A successful verification plan will provide stability for the verification team throughout the project. Most projects are somewhat dynamic in nature, not only due to architectural and design issues that arise during the project, but also due to external pressures that may affect the architecture or schedule requirements of the project.

At any time, a successful verification team should be able to understand which components of the system have been verified and what areas are still left to do, and have a current projection on the quality and completeness of the project overall. As portions of the project are modified, the verification plan must be updated so that the overall project impact is understood, and nothing is missed during the execution of the project. The verification plan can be used as a guide and a checklist throughout the project.

Phases of the Verification Plan

There are several phases that the verification plan must step through as it is being created. The order of the phases need not be exactly as is shown here, but some of them are dependent on others. There also tends to be some iteration as discoveries in one step may affect previous steps. The major reason for going through the phases is to ensure that the verification plan is complete and that the project schedules are aligned. While there is no set order, there are many issues that should be defined within the plan. Figure 8-1 shows one possible ordering of tasks within the development of the plan.

The first phase is to define the goals of the verification project. This may specify the required confidence level of the verification project, and provide guidelines for determining the completeness of verification. Establishing the guidelines first will ensure that the strategies for test writing and coverage metrics meet the overall goals of the verification plan.

Once the verification goals have been established, create a feature list to determine what needs to be verified. The feature list may range from high-level architectural features, to software use models, to the implementation details. Creating the feature list often requires significant input from various groups in the project. This will generally require at least some interaction with the hardware and software design groups, and potentially from the system architects. Since this is often a very early interaction between groups, it is not uncommon for bugs to be found and resolved during this review process.

With the feature list in place, plan the verification environment. The planning is often done from the bottom up for block-level testing, and then from the top down for system and subsystem environments. Once the initial requirements for the block and system environments are understood, common elements can be identified between the block- and system-level plans. Some modifications are generally required for

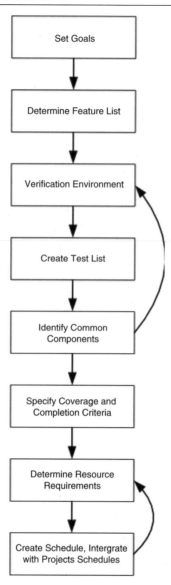

Figure 8-1: Phases of the Verification Plan

both the block- and system-level environment plans so that common components can be shared across environments. This is also the place where any applicable intellectual property from other sources should be identified and added to the verification plan.

Once the verification environment has been defined, the individual tests need to be listed and described. A test description will generally include the functions it is expected to cover, the components and configurations that must be in place for the test to be run, and an algorithmic description of the test if that will be needed.

Assertions that are to be used for verification should also be included here. While assertions are often created by the design team rather than the verification team, if they are to be considered a part of the overall verification strategy, then they must be included in the verification planning and reviewing process. The planning format for assertions is generally quite similar as for stimulation-based tests. This will include the functions to be covered, and the necessary components for the assertions to run. Assertion tests may also include information on where the assertions will be placed.

With the tests and features understood, identify components that may be needed by multiple tests. If a component can be shared by more than one test, then the verification effort can be reduced. The common components should be listed, and the verification plan may need to be updated to indicate where components are re-used rather than recreated.

For larger projects, a cross-reference of product features and tests may be useful to help in the determination of which tests are covering which features.

In addition to test requirements, the verification plan should identify the metrics that will be used to analyze the quality of the verification. This section should provide guidelines for determining when the minimum requirements for the completion of the verification phase have been met. This may include specific requirements for line and functional coverage metrics as well as others such as bug tracking metrics. The completion point is often referred to as tape-out. While tape-out usually implies a step in the chip fabrication process, the term is useful for other systems, such as those using FPGAs as the point where the system migrates from simulation to a physical implementation.

With all of the verification components and tests identified, there is usually sufficient information to determine the resources needed to implement the verification plan. This provides an estimate of the absolute amount of manpower needed to implement the environment and tests. These need to be meshed with the other project schedules, which should show when the components necessary to run each test are available. At this point, the schedules, budgets, and resource requirements of the verification plan need to fit with the other project components. All of this should be documented as part of the verification plan.

The Phases in Detail

Feature List

In some ways, this is the most critical component of the verification plan. This is a list of all the critical features of the system under test that must be verified. This will generally include architectural requirements of the system, the software requirements, as well as design features that must be verified.

This is often the area where a review with the system architects, and software and hardware engineering teams is necessary. The feature list needs to be sufficiently specific so that the verification team can have confidence that by verifying this list they have provided adequate coverage of the system. On the other hand, the features should not get too specific either. The tests should be created to test the functionality of the block, not the exact implementation of a specific feature.

The list may be prioritized with those sections that absolutely must function separated from any features that are desirable but not necessary. This prioritization may be needed as budget or time pressures require that the device be taped out before the verification is complete. The prioritization ensures that the most important functions are verified first, and may also provide a guideline for when an early tape-out is feasible.

In many cases, a significant fraction of the design effort is focused on performance goals. This is a part of the architectural feature set, and any required performance metrics or measurements should be included as a part of the project feature list.

The Verification Environment

With the key features understood, a determination of how these features will be tested needs to be made. Many features are easiest to test at a block level first in order to ensure correct operation, and may then need to be retested at a higher level to ensure interoperability with other components.

The verification environment must be designed to support the appropriate block-level tests as well as the migration up through subsystem- and system-level tests, and to facilitate sharing between the various subcomponents of the system. Configurations are often used for this.

Configurations

Configurations are often used to specify the level at which testing is to be run. Smaller block-level tests are run in a configuration that allows just the necessary part of the verification environment to be used. Larger configurations will put multiple blocks together to be tested in a multi-block- or system-level environment.

Verification tends to start at a block level first, since blocks are often easier to debug alone. As blocks are integrated, more configurations are needed to support multi-block simulations. Figure 8-2 shows a sample block-level configuration.

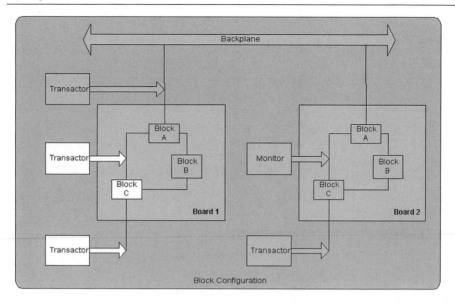

Figure 8-2: Block-level Configuration

As the block is integrated into the larger system, multiple blocks may be connected together. These examples are simplified, since there are generally many more blocks than are being shown. Figure 8-3 shows how multiple blocks may be integrated into a subsystem-level simulation. Transactors that previously drove an internal bus may now be used to monitor the bus as multiple blocks communicate with each other.

Generally, there are some simulations that will be run on a full system configuration. Depending on the size of the system, the number of transactions run may be limited by the speed of a large simulation. Figure 8-4 shows the progression of configurations as blocks of hardware are incorporated. Clearly, software blocks may also be used in each of the configurations.

Verification Components Planned for Various Configurations

It is important to understand the required configurations at the start of the verification process, since the verification environment will wrap

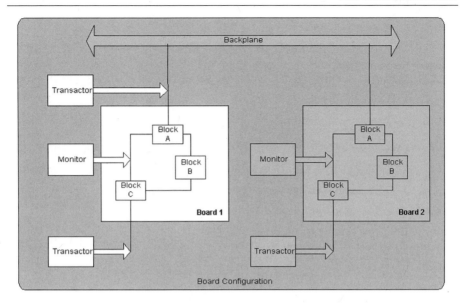

Figure 8-3: Subsystem-level Configuration

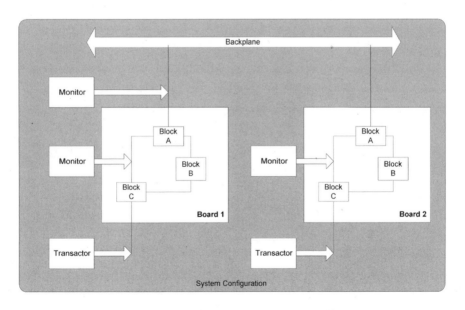

Figure 8-4: System-level Configuration

around each configuration. This is the first area where components may be shared. In the example shown in Figure 8-2, the transactors used in various device configurations may also be used in subsystem- and system-level configurations.

When choosing configurations, it is important to find a good balance between the number of configurations and the requirements of the test plan. Since configurations take time to build and debug, they are not cost-free. Too many configurations will cost project time. On the other hand, a configuration can save in test effort, since a test will run faster, and may have more direct control over a particular block of logic. Choosing an effective subset of possible simulation configurations tends to minimize the overall functional verification time and cost.

Re-use of Verification Components

During the writing of the verification plan is often when the verification environment structure is first being determined. This is when the components of both the block- and system-level tests can be identified. It is also in this stage that verification re-use can be planned. First, the common components such as transactors, drivers, or test modules can be identified at this point. Block-level tests often require transactors or models that will also be needed in larger environments. By identifying the shared components early, the verification environment can be structured to maximize the shared components. This may be a somewhat iterative process. Some environments may need similar components that are not identical. In some cases, it may be worth redesigning the environment so that a single component can be shared. Re-use does require some management overhead to ensure that the blocks will be compatible and are written in a modular fashion.

The verification planning can also incorporate any pre-existing components from other sources. When re-using components from other projects or from third-party sources, it is important from a scheduling viewpoint to have some understanding of the quality and support structure around that component. There should be some understanding of the operation of the component, and how any issues are likely to be resolved. If there is no support available, then some time should be scheduled for understanding and maintaining the component.

Test List

While the test list and the verification environment often do go hand-in-hand, the complete test list is usually finished last, since there is often a lot of detail in the test list. Once the structure of the verification environment has been sufficiently defined, the test list can be defined in detail. Before this point, there is usually a rough outline of the test list that is used to help define the required verification environment. There are often a few iterations to ensure that the verification environment can support the tests that are deemed necessary.

The detailed test list is simply a complete list of all the tests that should be necessary to verify the functionality specified in the feature list. A good test list should provide more detail to ensure efficient execution of the test plan.

Features

For each test in the list, several other pieces of information are generally specified. If a set of assertions will be used in the verification plan, then the location and method of the assertions should be specified. For simulation tests, the necessary configurations and components should be listed. The list of required resources will allow each test to be integrated into the project schedule based on when the necessary components are available for the test to be run.

For both assertion and simulation-based tests, specify a list of features that each test is expected to cover. This information is usually added to a cross-reference table, often implemented as a spreadsheet, that correlates the feature list with the test list. This table can be used to determine which tests cover a feature and vice-versa. It will also show if a feature is covered multiple times. Having this list is important not only to confirm that the test list is complete, but is often a reference later in the verification project if any issues arise during the coverage analysis phase. Should the analysis show that sections of the design are not sufficiently covered, this cross-reference can be useful to determine where the weaknesses occurred. This could mean any combination of a missing feature, a poorly defined test, or a test that was not implemented according to the specification. Understanding the source of any coverage weaknesses may help in finding other unknown problems.

Descriptions

All but the simplest tests in the list should also include an algorithmic description of the test and the checks that will be made. This is particularly important if the verification plan is to be implemented by a larger or more diverse group of verification engineers. With a specific test algorithm defined, there is less room for deviation between the test specification and the implementation. A good test description may also be important if the verification tests are to be spread out to a large group of people. By understanding the test requirements and algorithm, as well as the checking methods, it is usually possible to determine the complexity of each test and ensure that each test is assigned to an engineer with the appropriate skill level for that test.

Coverage and Completion Requirements

One of the well-known dilemmas of verification is determining when the verification effort is done. Since it is never possible to determine if all bugs have been found, multiple techniques are used to estimate the level of coverage that has been achieved.

Tape-out Decision

Since all of the coverage metrics are estimates, a difficult decision will arise as the verification effort nears the end. This is determining when the tape-out should occur. There are almost always significant pressures to tape-out as early as possible, due to resource or schedule requirements. On the other hand, there is almost always more verification that could be done, and some uncertainty about the quality of the verification effort. Because of these competing pressures, it can be quite difficult to make a tape-out decision that balances the schedule requirements with the verification confidence-level requirements.

To provide guidance for the tape-out decision, the verification plan is a reasonable place to state the expected completion criteria of the verification project. During this phase of the project, the verification plan is being created to ensure that the coverage will meet the defined goals. This is an ideal time to state the completion requirements of the verification effort.

Completion Components

The completion requirements will have several components. These may focus on test coverage, coverage metrics, or bug metrics.

It is generally easy to define more tests in the verification plan than the project has time to complete. When this is the case, it is important not only to prioritize the test list, but also to state which tests are viewed as essential to complete before tape-out.

In some projects, a test review must be done at the end of the verification project as well. This type of a review is intended to re-confirm the functional coverage requirements and the test outline for each test and to ensure that the features have indeed been covered. Because features and tests almost always are added during the course of a project, this last review may help ensure that the modified verification plan is still complete. As before, this review requires that the appropriate people from various groups be involved.

Coverage metrics may provide a significant amount of information about the verification test suite as well. Most projects will specify a certain coverage goal that should be met, for example 85% branch coverage. While this may be a good goal for the project, it is quite vague. There may be some areas that have significantly lower coverage measurements, even though the average meets the target completion requirements. Deciding if that is acceptable, or whether each component should meet the coverage target, should be determined early. This may be important, since a single area of low coverage could indicate a weakness in the verification plan, even though the overall metrics indicate that the verification coverage is good.

Coverage goals should also specify which types of metrics are to be used. Functional coverage measurements provide a different view of the verification quality. By using a variety of coverage metrics, a better overall view of the verification quality can be obtained.

Finally, completion requirements may also use bug-reporting metrics as a way of estimating the quality and quantity of the verification effort. Tracking bug reports as the project progresses usually does this. The reports should record which section of code the bug was found in, which test discovered the bug, and perhaps which group the particular test belonged to. By taking both the number of bugs and the size of the design into account, the bug reports will provide a picture that shows the density of bugs in the design code as the project progresses.

Many groups have found that bug densities are relatively constant for code of similar complexities and design styles. As a result, if the bug reports indicate that some sections of code have abnormal bug densities, either too high or too low, then further examination of those sections may be called for.

Along with bug densities in the design code, the bug reporting can also track the density of bugs found by the verification code. Individual tests are not likely to find similar numbers of bugs simply because of overlap between tests. A test that is run earlier in the project cycle is likely to find more bugs than one that is run later. However, groups of tests that cover a particular feature set are generally expected to find some number of bugs within that feature set. A lack of bugs may indicate a problem with the verification approach to that particular feature.

The completion requirements may call for a review of bug densities, along with a follow-up of any suspicious findings as a part of the tape-out decision process.

It should be noted that for bug reporting to be accurate, the reports must start as soon as any significant verification effort begins, and that all bugs must be accurately recorded. Bug reporting should be viewed as a normal part of the verification effort, and should generally not be used to judge the quality of any section of code or any engineer. A larger than normal number of bugs is generally not an indication of problems with the code. The normal indications of problems are bugs that are opened but not resolved in a timely fashion.

Resource Estimation

At this point in the verification plan, a great deal is understood about the verification project. The environment, configurations, and tests have all been itemized, and it is possible to determine how much effort will be needed to implement the verification project.

In estimating the resource requirements, it is important to list not only the expected time required to complete a particular task, but also the skills and level of expertise needed for each task. This will provide the information necessary to determine which group or engineer might be suited to any particular task.

Along with the time and complexity estimates, a list of dependencies for each task will be useful. This is simply a list of all items that must be completed before any task can be executed. This information will be needed for the scheduling of the verification project.

Project Schedule Integration

If the verification plan has been done carefully, integrating the verification schedule with the rest of the project schedule at this point has very little to do with verification. This becomes a classic project management task that must balance the tasks, dependencies, schedule, and resource requirements of the entire project, not just the verification portion. If the previous sections have been documented correctly, then there is sufficient information for a project manager to create and integrate the schedules.

In project scheduling, it is important to allocate time for block integration and code debugging. If the verification group is expected to integrate the blocks into subsystem- and system-level environments, and provide effort for the debugging of design code issues, then these efforts must be accounted for in the project schedules.

During this task, an accurate determination can be made about the size and makeup of the verification group that will be required, as well as the computational and software requirements. The project scheduling will be the point at which the time and budget resources tradeoffs are made. As a result, accurate scheduling ensures that the corporate resources and marketing requirements are all met.
A thorough verification plan can help ensure that these estimates are as accurate as possible.

Summary

The verification plan is a large document or set of documents with many detailed sections. It requires input from many groups within the project, and it has sections that can be extremely detailed and time-consuming to create. As a result, there is sometimes a tendency to try to skip or shorten sections of this document.

Because the verification plan is a critical component of the overall project flow, it is important to ensure that the plan is complete and thorough. The verification plan is often a central component to the organization of the verification project, and critical to creating an accurate estimate of the time and cost requirements of the project.

A solid verification plan will not only provide a guide to the scheduling effort, but will also ensure that the system being designed is verified in an optimal fashion within the resource constraints of the project. This in turn maximizes the likelihood of a functional system when the tape-out decision is made.

Applying Functional Verification to a Project

Key Objectives

- Understanding project costs

- Tool requirements

- Project management

The start of a verification project can be overwhelming. There are an enormous number of issues to address and decisions that need to be made. These will include determining the requirements and goals of the verification project, the available budget and team, and starting the specification of the verification plan, methodology, and tools. The major constraints are generally related to time, budget, and the availability of existing teams and software. From the general constraints, decisions must be made in terms of tools, methodology, licenses, team size, and project schedules as well as many other issues. Most of these issues are tightly interrelated making it difficult to separate any one item out.

The consequences of some decisions will be far reaching. Others will be of little importance. Keep in mind there is no single right way to perform functional verification. Rather, there are a number of methods and tools available, and an efficient approach will match the right tools with each phase of the project's needs. For most projects, a combination of tools will be useful.

Organizing the requirements of the project, and determining the appropriate verification tools and methods based on those requirements are often reasonable ways to start the verification project.

Since cost is often an overriding factor, it may be important to determine where project costs can come from. While this may vary greatly for any particular company or project, listing the cost sensitivities may help in making decisions.

Project Costs

For many projects, the project costs may be sorted roughly in this order, from most expensive to least expensive:

1. Time

2. Personnel

3. Licenses

4. Equipment

Understanding the relative expenses of each of these may affect decisions in the planning of functional verification as well as other parts of project planning.

Time

This is often the single most expensive variable of a project. The time during which a project is being developed carries significant costs for a variety of reasons. Until a new product is developed it cannot be sold. Any increase in the product development cycle has a corresponding decrease of the product sales cycle. This is amplified by the fact that the product is likely to be most valuable during the early part of its sales cycle. As a product matures, competing products often become available that are faster or cheaper, and thus reduce the profit of the more mature product.

A product that is developed faster will require less corporate overhead, and is likely to result in a more profitable product. While a good product methodology can reduce a project cycle somewhat, there is often a tradeoff between speed and project cost. Cutting a project schedule time in half will generally require more than twice as many people due to the overheads of training, coordinating, and communicating between a larger group.

Since time is such a critical factor, pressure is often placed on engineering teams to reduce the product cycle. While this is a laudable goal, many projects have actually taken more time to complete by cutting corners. A successful project must be aware of time costs and balance the schedule time with project risk to reduce the overall project time.

Personnel

After time, this is usually the second most important cost consideration. Engineers are expensive in terms of salary, but also for the overhead they require. For example, when evaluating a new tool, it may be important to consider training costs. A typical training cycle will often require a week of formal training, plus one to six months of hands-on experience to become proficient. When that cost is multiplied by the size of an engineering team, the cost can be quite substantial.

Licenses

Tool licenses can look expensive. The numbers are big, and most tools are licensed in such a way that quite a few licenses are required. The result is that tools show up as a big expense early in a project. As a result, this is sometimes an area where cuts are made.

In many cases, the licenses costs are actually small when compared to time and personnel costs. As a result, this may not be a great place to save money. It is not uncommon to see projects where engineers are waiting for access to licenses. This may not be a good way to minimize overall costs, since it is likely to increase the project schedule in order to minimize license costs. In addition, the engineers are on payroll, waiting.

Licenses can be managed, however, since most projects will start with fairly low license usage when the project is still small. As the size of the design grows, and verification tests become more numerous and complex, the license usage will increase. It may be possible to spread the license requirements to reduce the costs somewhat without affecting the overall project time.

Equipment

Equipment costs have been dropping so fast that this rarely even looks expensive anymore. However, the costs in equipment are more complex than just ensuring that there are sufficient numbers of servers. A bottleneck anywhere in the system can severely affect the speed of simulations. In addition to ensuring that there is a fast processor with sufficient memory for each simulation, bottlenecks can show up in the communication to storage devices, excessive output to display devices during simulation, or a variety of other places. Measuring the simulation effectiveness and fixing problems detected at various times during the project will ensure that inefficiencies are found and corrected early.

Methodology

Many possible methodologies to functional verification have been discussed. For any particular project, only some of them will be needed. Understanding what will work best for a project, and deciding on the methodology, should be done early. A successful project is likely to have a clearly defined methodology that is well integrated with the tools to be used. A consideration to remember when choosing is that any new methodology that is used will require training for the engineering team.

Requirements for a Project

The first part to deciding on a methodology is to understand the project requirements. This will include factors such as the complexity of the project, the number of components that are being re-used from previous projects (and are therefore at least partially tested) or that are being designed from scratch, as well as the time and risk requirements.

This is where some of the basic goals of the verification project may be defined, such as the time and risk requirements, which will specify the more specific tape-out criteria for the project. With the high-level requirements in place, and a methodology chosen, the task of creating the verification plan can begin.

Tools

The choice of tools is critical to the success of a project. The most important issue is that the tools work well with the methodology that will be used. Since tools and methodology are interdependent, it is generally a good idea to choose the methodology and tools at the same time. Choosing one without understanding the requirements of the other often results in a less-than-ideal project flow. Beyond that, making sure that the engineering team is trained and comfortable with the tools, and that the required support is in place, are also important.

Types of Tools

There are many different tools available for functional verification, most of which have been discussed in previous chapters. Because tools are often bundled together in various ways, it may not be cost-effective to select each tool individually. With a good understanding of the project requirements for each type of tool, it should be possible to determine if any particular bundle of tools will meet the project needs.

Simulation and Assertions

The simulator is probably the single most critical tool. There are multiple factors to consider here, including the desired simulation language or group of languages, the performance requirements, and the price. Simulators have been around for a long time, and as a result, the requirements are well understood, and the products have matured. This means that the simulators are stable and competitive.

Along with simulation, if dynamic or static assertions are to be used, then this may require an additional tool, or one or both may be bundled in with a simulator. Assertion tools can be powerful, but they are not as mature a product as simulators.

Verification Languages

If a verification language is to be used, there are many tools available in this area. Many of the concepts are similar between the verification tools, and many verification tools will bundle in one or more analysis tools.

Despite some claims to the contrary, one can successfully verify a design with any of the major verification tools. It may take longer and be more difficult if the tool does not fit well with the other tools, or the methodology, or if the team is not comfortable with the capabilities of the tool. Simulation speed and cost may also be a factor.

Linter

Probably the first analysis tool to be used in any project is a linter. This tool examines the RTL code as soon as it is written. This is when the designer still understands the details of the code, and no other group is relying on the code yet. A good linter will look for a wide variety of questionable design practices in RTL code, and will provide a set of warnings. Generally it is a good idea to use a linter and follow the recommendations that it provides. It is also a good idea to develop and use a coding style guide for all code that is developed in a project, not just the design RTL. A linting tool tends to be a good investment, since it finds and reports bugs early that would otherwise require debug time to find.

Coverage Analysis

As was discussed in Chapter 7, there are a variety of coverage analysis techniques, and quite a few tools available that will perform most or all of the coverage methods. In many cases, a coverage tool is bundled with a simulator or a verification tool.

Functional coverage and code coverage tools are quite different, although several tool vendors bundle them together. It is important to understand the project requirements for coverage analysis, and that the chosen set of tools will meet those requirements.

Waveform Viewer

A waveform viewer can greatly ease the debug of low-level design issues. While waveform viewers traditionally just displayed signals in the RTL, some are starting to incorporate data from the verification language, software code, and transaction-level information.

Most simulators have a bundled waveform viewer available, or one can license a separate viewer. While viewers are not always the best tool to debug problems, since they can significantly slow the simulations, they are often ideal for low-level debug, and can provide a degree of confidence to an engineering team. A good tool choice is one that is able to display information from the various languages that will be used, and that the team is comfortable with.

Results Analysis

A results analysis tool is often a useful way to determine how well a system performed. These tools may use the transaction database discussed in Chapter 4 to analyze overall system performance, or utilization of various components, and to display that information. A good analysis tool should be able to gather any type of system information from the transaction database and allow a variety of analysis and display options.

Where system performance is critical, and design tradeoffs are being made in order to meet architectural and cost requirements, a results analysis tool may be useful to help understand how well a system is performing, and where the bottlenecks are.

Tool Evaluation

When evaluating tools, it may be important to keep the basic principles of functional verification in mind. As with any product, the claims from some vendors are likely to be somewhat exaggerated. Verification tools have a special challenge, since there is a long time lag between when a tool is purchased, and when actual results can be measured. Measuring the effectiveness of functional verification is difficult until a full project cycle is complete, and a design has been implemented and is running in real customer sites. Only with the experience of the actual design running in the intended environment can it be determined how many bugs were found, and an estimate be made of how many made it past the functional verification effort. This delay, and the fact that there are no controlled experiments, makes it difficult to evaluate the effectiveness of any particular tool or approach. Unfortunately, this might also encourage a tool vendor to make claims that would be difficult to substantiate in other circumstances. It is also possible that tool vendors themselves don't really know, since they too suffer from the same time lag from tool development to results.

As a result, it is important to understand the functional verification process and principles when evaluating tools, and when deciding on the tools and methods to use. Whenever tools are described that will automate the process, make a complex task easy, or make all the problems go away, it is time to ask many very detailed questions to understand how that would work. Simple demonstrations are very different from the complexities of an actual project.

Examine the usage model for the tool. By understanding what the inputs are, the types of interactions that will be needed, and how the tool goes about creating the outputs, it is often possible to understand what the tool will do for you, and just as important, what it won't be able to do for you. This will also help in understanding how the tool will fit into the overall methodology, and where it must interact with other tools.

Keep It Simple

There are a number of tools available on the market that attempt to use new techniques or that promise dramatically better results. While such tools should generally be investigated, there is sometimes a difference between finding a new way to solve a very specific problem and finding the optimal complete set of tools necessary to solve the problems of a specific project.

Simple approaches are often better, since they are easier to understand, easier to train engineers on, and easier to debug when things go wrong. If the simple approach will provide the same benefit as a more complex approach, it is always best to stick with the simple approach. New tools also carry an element of risk, since they may be unknown to the engineering team.

The trick to this advice, of course, is to recognize when there is real value to be gained by a more complex approach. As always, there can be a lot of time between when a decision is made and when the results of the decision will be understood. As a result, any real data may be helpful. Looking for real-world data on a tool or methodology, rather than relying on the results of a simple demo is likely to aid in making a good decision. Any tool can be made to look good in a demo.

A design with many millions of gates is likely to find the pitfalls that the demo didn't show.

Structuring a Project

There are a variety of tasks that must be performed in a functional verification project. Some of these are complex, while others are more self-contained. Making sure that the skills of engineers are in line with the requirements of a task is important. Higher-complexity tasks require a greater skill level. Unfortunately, if the skills are not properly aligned, then the tasks may not be completed, or will cause problems as a particular component is integrated into a larger verification environment.

Identify Components

The Verification Environment

The creation of the verification environment, and the configurations that will be needed, form one of the more complex tasks in the functional verification effort. This area requires knowledge of the product architecture and the verification architecture, as well as the tools and methods that will be used.

Because the environment is created before it is heavily used, there is little feedback on the correctness of the environment until after a great deal of work has already been done. As a result, problems in the architecture or design of the verification environment may have a significant schedule impact.

Because of this, it is the most experienced people on the verification team who handle the architecture and design of the verification environment. Frequent detailed reviews are also a good idea to ensure that nothing is being missed. While the environment is one of the most complex components to build, it must be easy to use, since the entire verification team will be working within the environment.

Transactors

After the environment construction, the design of transactors is probably the second most complex task. Building a transactor requires detailed knowledge of the specific protocol. This is not just the basic operation; there also needs to be an understanding of all possible error or retry conditions, and the range of legal responses to each condition. A transactor is expected to push the limits of any particular protocol, and thus the transactor designer must understand what those limits are.

In addition to understanding the protocol, the transactor designer must also be able to create a driver that provides the tests with a clean interface to create and receive the transactions. As with the environment creation, this task requires a solid understanding of the verification methods, environment requirements, and test requirements. This requires that the transactor designer have a broad understanding of all of these. Again, this task is generally reserved for highly experienced engineers.

Support Routines

Support routines may be drivers or software implementations of various algorithms that will be used in the verification code. An example of a support routine is a checksum generator that will create a valid checksum value for generating or checking packets.

Since support routines are self-contained pieces of code, they tend to require knowledge of the algorithm itself and the language in which they are written. As a result, they do not require the experience level of other tasks. These are good candidates for engineers that are not as familiar with verification, or that are new to the project.

Test Writing

Test writing is often the single biggest task of a verification project. Test writing in a well-designed and structured verification environment should be relatively straightforward. The primary requirements to write a test are an understanding of the architecture that is being tested, an understanding of the verification language, and a user-level view of the verification environment and transactors.

While some tests are highly complex, others will be focused on specific areas of the design, and are easier to create. As a result, test writing does not generally require very advanced verification or tool knowledge, although a good test writer must understand not only how to generate specific conditions, but also what one can and should check. With a good test plan in place, test writing is a place where new engineers can be effectively brought into the middle of a project. While most tasks require many months of background learning that come from being involved in a project from the start, it is often possible for an engineer to become productive in test writing much sooner, since it is possible to limit the amount of information needed to write a test.

One of the other simplifications to test writing is that there tends to be immediate feedback. As soon as a test is written it can be run against the design. As a result, any problems with the test will be found soon after the test is written. This makes the task of test writing significantly easier.

Project-level Debugging

Verification tests are intended to find and report errors in a design. When an error is reported, it may be a design error, but it may also be an error in the test, the specification, the transactor, or the verification environment.

In some cases, determining where the error is, or if the response is correct, can be a complex task. Generally there are a few people in a project that understand all of the pieces, and how to determine where a problem is. The engineers that understand the whole flow will find that they spend a great deal of time helping other team members, and less time writing tests themselves. This is often an efficient use of their time, since it leverages other team members.

Understanding who these engineers are, and ensuring that they have the time to help the rest of the team is often in the best interests of the overall project flow. Because these engineers will be working on problems across the verification project, their tasks can be hard to schedule.

Flexible Manpower Requirements

In addition to ensuring that the engineering skills are matched to the tasks, a well-run project will often have a variable number of tasks that must be performed during different phases of the project.

At the start of a project, a great deal of code is being developed, but not much is available for functional verification for a while. This is the time that the verification team is creating the transactors and integrating them in a new environment. However, since there is often not much RTL code available yet, the test writing is not yet going as fast as it will.

As the RTL code becomes available, the test writing effort increases. Early in the project, few tests are generally able to find many bugs. As the quality of code increases, more tests are required to find fewer bugs. At this point, there are generally a large number of tests to be written, and this is often the single largest effort in the project.

As a result, many projects need a few very senior verification engineers at the start of a project. As the project progresses, more engineers are needed to help in the test writing phase. At the peak of the project, the group size may be several times larger than the starting team. As the majority of tests are written—and passing—and the test quality level is shown to be sufficient, then the size of the team can be reduced as the last tests are completed.

Using a structured verification environment, it is possible to bring additional people into the verification effort as they are needed. For an engineer unfamiliar with the project to be effective, it is important that the environment, tasks, and design be well documented so that people can learn what they need to know quickly. While this requires some overhead, the flexible manpower can be used to accelerate the project schedule while keeping costs in control. In many cases, cutting the project time in half will more than double the project cost. This is because a larger team requires more communications, and a greater number of smaller tasks to manage and interleave. This can still be worthwhile for an earlier product completion.

Project Management

The management of a verification project is reasonably similar to any other technical project management. There are some areas where there may be differences.

One of the big challenges to managing a verification project is that there is often a significant gap between the time that a task is completed and when the quality level of that task can be measured.

An example of this is the entire verification flow of a project. Between the time that the verification effort is completed and the device is taped out and tested in the lab may be several months. Only after the real device has been run in a real system is there any definitive feedback on the quality of the verification effort. By the time that feedback is available, it is usually too late to fix anything without a significant schedule impact and cost.

Because of this lack of easily available quality metrics, it is important to put additional effort into understanding both the progress and quality of the verification effort from the start. By measuring frequently and accurately, it is usually possible to find a problem early enough that it can be fixed without drastically affecting the project schedule.

Understanding Project Progress

One of the challenges to ensuring project success is to detect problems as early as possible. In this way, time is not lost, and there are generally more options available to correct any problems. This holds true for verification projects just like any other projects.

To help provide insight into the quality and progress of the verification effort, it is important to have specific criteria that can be used to measure and monitor both the quality and progress of the verification effort. If the verification code is written, and is waiting for a design to become available before it is run, then one can measure progress, but not quality. Without running the code against a design, it is unlikely that the code has been debugged, and it is quite difficult to estimate the quality level. As a result, from a project management viewpoint, there is significant uncertainty about the current state and progress of the project. Finding specific criteria to measure and monitor progress is important to monitoring project progress.

There are various ways to determine how well individual tasks are progressing. In verification, completing the writing of a test is of less interest than determining the effectiveness of the test. Careful bug tracking will show if tests are finding sufficient bugs, and if the bugs are being found in all sections of the design. Similarly, the bug open rate, the rate at which new bugs are being found, may be a good indicator of test quality. The bug close rate, the rate at which bugs are being fixed, can point to potential problems such as an issue with the quality of the design RTL. The frequency of new releases may indicate an issue with the verification environment. On the other hand, any of these indicators may simply be indicating a great deal of activity.

Examining and tracking all activity can provide clues to problems in the entire verification effort. What is important is to look at all available information to find trends that could indicate that an issue exists somewhere in the process. It is important to watch the activity, since too frequently problems go unnoticed for too long and affect the overall project schedule.

Summary

Functional verification is not straightforward, and as of yet there is no single, agreed-upon approach to verifying a project. However, with careful planning and management, and a good understanding of the goals, methods, and tools to be used, a functional verification effort can run smoothly, and can meet the cost and risk requirements of the project.

Additional Reading

Verilog
Advanced Digital Design with the Verilog HDL. M. Ciletti, Prentice Hall, 2002.

Verilog HDL. S. Palnitkar, Prentice Hall, 1996.

VHDL
VHDL Coding Styles and Methodologies. B. Cohen, Kluwer Academic Publishers, 1999.

VHDL Starter's Guide. S. Yalamanchili, Prentice Hall, 1998.

RTL Coding
Principles of Verifiable RTL Design. L. Bening and H. Foster, Kluwer Academic Publishers, 2000.

Reuse Methodology Manual for System on a Chip Designs. M. Keating and P. Bricaud, Kluwer Academic Publishers, 3rd edition, 2002.

Verification Languages
A SystemC Primer. J. Bhasker, Star Galaxy Publishing, 2002.

System Design with SystemC. T. Grotker, S. Liao, G. Martin, and S. Swan, Kluwer Academic Publishers, 2002.

Vera: The Art of Verification with Vera. F. Haque, J. Michelson, and K. Khan, Verification Central, 2001.

Formal Verification

Computer-Aided Verification of Coordinating Processes. R. Kurshan, Princeton University Press, 1995.

Model Checking. O. Grumberg, E. Clarke, and D. Peled, MIT Press, 1999.

Symbolic Model Checking. K. McMillan, Kluwer Academic Publishers, 1993.

Other Verification Resources

Writing Testbenches. J. Bergeron, Kluwer Academic Publishers, 2nd edition, 2003.

Index

NOTE   *Those entries shown in italics indicate figures*